CW00554251

THE OXFORD CHEKHOV

VOLUME I

SHORT PLAYS

THE OXFORD CHEKHOV

VOLUME I

SHORT PLAYS

Translated and edited by

RONALD HINGLEY

OXFORD UNIVERSITY PRESS

OXFORD NEW YORK TORONTO MELBOURNE

Oxford University Press, Walton Street, Oxford OX2 6DP
OXFORD LONDON GLASGOW
NEW YORK TORONTO MELBOURNE WELLINGTON
KUALA LUMPUR SINGAPORE JAKARTA HONG KONG TOKYO
DELHI BOMBAY CALCUTTA MADRAS KARACHI
NAIROBI DAR ES SALAAM CAPE TOWN

ISBN 0 19 211349 6

First published 1968
Reprinted 1979

Printed in Great Britain
at the University Press, Oxford
by Eric Buckley
Printer to the University

To

ALEXANDER KERENSKY

CONTENTS

PREFACE

(*a*) *Contents of this volume*

THIS volume contains all Chekhov's extant short dramatic work, including all variants except for certain trivial items which could not be reflected in translation without pedantry, and one or two other minor omissions which are indicated in the relevant appendixes. Thus volumes i to iii of this edition give, for the first time in English translation, Chekhov's entire dramatic *œuvre* presented in a rational order.

The short plays are printed (as are the long plays in vols. ii and iii) in what is basically their chronological sequence, but with certain minor divergences dictated by special circumstances. In the determination of the volume's exact contents and of their sequence some minor problems have arisen. In particular, do the differences between the play entitled *Smoking is Bad for You* as published (*a*) in 1889 and (*b*) in 1903 entitle one to regard this as two separate works?[1] Or should the former version be regarded as a draft only? The second solution has been adopted here.

(*b*) *The text*

The translations in this volume are based on the Russian text as printed in vols. xi and xii of the twenty-volume *Complete Collection of the Works and Letters of A. P. Chekhov* (Moscow, 1944–51). This edition is here referred to as '*Works*, 1944–51'.[2]

(*c*) *Treatment of names of persons*

The treatment of names of persons follows the lines laid down at some length in the prefaces to vols. iii and viii of the present edition.[3] Briefly, the aim has been to convey the relationships between the characters by stylistic means rather than by mechanically reproducing the names as they occur in the Russian text. Fairly free use has been made of English Christian names where these seem appropriate, i.e. in the following instances:

[1] It is so treated in *Works*, 1944–51, xi; see further, Appendix ix, below.

[2] For the full Russian title of this work, see Select Bibliography (below), p. 209; for a longer note on *Works*, 1944–51, see Preface to vol. iii of the present edition, p. ix.

[3] See *The Oxford Chekhov*, vol. iii, pp. xi–xvii; vol. viii, pp. ix–x.

Anglicized form	*Russian form*	*Anglicized form*	*Russian form*
Alexander	Aleksandr	Marcellus	Markel
Alexis	Aleksey	Mary	Marya
Andrew	Andrey	Nicholas	Nikolay
Barbara	Varvara	Peter	Petrushka Pyotr
Constantine	Konstantin		
Epaminondas	Epaminond	Procopius	Prokopy
George	Yegorka	Simon	Semyon
Gregory	Grigory	Stephen	Stepan
Helen	Yelena	Theophilus	Feofil
Luke	Luka	Veronica	Veronika

In *The Night before the Trial* the name Fedya is rendered 'Fred'.

(*d*) *Aims of the translation and special problems*

The chief aim has been to produce versions for the stage in accordance with the general policy described in the Preface to vol. iii of this edition.[1]

Certain specific problems are also touched on in other volumes published earlier than vol. i: the rendering of uneducated Russian speech, and also of the (very frequent) terms of abuse.[2] The same policies have been followed in the present volume, where an additional problem is posed by the numerous terms of endearment which form such a characteristic feature of Chubukov's speech in *The Proposal*. It has been decided, contrary to normal policy in this edition, to give each of his locutions a single unvarying English equivalent ('old bean', 'old horse', etc.), thus preserving an essential feature in the stylistic pattern of the Russian.

Another special problem was posed by the rendering of the language of the Orthodox marriage service which figures so prominently in *Tatyana Repin*. It should be appreciated that the original language of these liturgical passages is not Russian but the cognate tongue Church Slavonic, which is employed by the Russian Orthodox Church in its ritual. For these passages, which repeatedly dovetail with dialogue couched in ordinary conversational Russian, I decided not to make my own translation but to employ instead the nineteenth-century English

[1] See *The Oxford Chekhov*, vol. iii, p. x.
[2] Ibid., vol. viii, pp. xi–xii; vol. ii, p. xi–xii..

rendering by G. V. Shann. Readers familiar with Anglican ritual may find this somewhat strange in flavour with its 'cedars of Libanus' (for 'Lebanon'), 'unbeginning Father' and so on. But I decided not to emend such locutions, feeling it appropriate to use, unchanged, an early translation in which the stateliness of Church Slavonic is, in my opinion, most effectively preserved. The tempting idea of concocting my own parody of the Anglican marriage service was thus successfully resisted. Incidentally, if any reader should decide on a Russian Orthodox marriage in England, he may well find his union blessed in the Shann version, though there are also several other accepted English versions.

It was stated in the Preface to vol. iii that 'An attempt has been made to use modern English which is lively without being slangy'. Owing to the character of the material in the present volume—which includes so much uproarious farce—more deviations in the direction of slang have been admitted than in other volumes so far published, as also a slightly greater degree of freedom in departing from the literal meaning of the Russian. If farces of this kind are not suitably 'racy', they are of little use for the stage. And there are many instances here where a too literal translation is disastrous, except for that (dwindling?) number of devotees who must at all costs have their Chekhov served up quaint. These still have (to take an extreme example) their Garnett, e.g.: 'It is easier to find a cat with horns or a white snipe than a constant woman',[1] where the present translation has 'You'll never find a constant woman, not in a month of Sundays you won't, not once in a blue moon!' (see p. 58 below). Similar deviations from strictly literal translation have sometimes been admitted where Chekhov employs a proper name familiar in his own day but possessing associations which may be lost on modern audiences without ready access to an encylopaedia. Thus, I have 'He doesn't half fancy himself, I must say!' (p. 137), where Miss Fen has 'He seems to fancy himself as a sort of Gambetta!';[2] contrast also with Garnett's 'not a general, but a peach, a Boulanger!'[3] my 'He's a General and a half, a regular conquering hero!' (p. 126).

In order to assist the flow of translation, шафер has been rendered 'best man', despite the fact that there would be more than one of them at a Russian wedding. For similar reasons, in *The Night before the Trial*, почтовая станция is sometimes rendered 'coaching inn', and its

[1] *The Cherry Orchard and Other Plays*, tr. Constance Garnett (London, 1950), p. 244.
[2] *Plays by Anton Chekhov*, tr. Elisaveta Fen (Harmondsworth, 1960), p. 427.
[3] *Three Sisters and Other Plays*, tr. Constance Garnett (London, 1935), p. 289.

manager 'innkeeper' or 'landlord'—in preference to 'post station' and 'station inspector'.

(*e*) *Acknowledgements*

It is a great privilege to acknowledge the help of the Archimandrite Kyril of the Moscow Patriarchal Diocese in England, who most kindly advised me on the relative merits of English versions of the Orthodox Liturgy. It was on his recommendation that I chose G. V. Shann's version to incorporate in the translation of *Tatyana Repin*; he also very kindly lent me a copy of this rare volume. Warmest thanks are also due to my pupil Miss J. V. Balmer, to Mr. H. S. Hayes and to my neighbour Mr. John Ridehalgh, who all very kindly helped to steer me through the difficulties of translating the sailing terminology in *The Wedding*; but if this material has come adrift at any point, the reader should of course blame my seamanship, not theirs. It is also a pleasure once again to thank Mr. I. P. Foote for performing the inestimable service of reading and criticizing the draft translations in typescript; as also Mrs. Olga Bowditch for her assistance with typing and my wife for her constant help and advice.

RONALD HINGLEY

Frilford, Abingdon
1967

INTRODUCTION

In bulk of output and—arguably—in literary achievement, Chekhov was first and foremost a short-story writer, and only secondarily a dramatist. And since his reputation as a dramatist happens to rest largely on the last four full-length plays from *The Seagull* to *The Cherry Orchard*, the ten short items in this volume occupy, inside the whole territory of his achievement as a writer, a minor enclave within a minor enclave. But they remain much more than inconsiderable trifles, many being in constant demand as one-act plays, justly famed on the Russian and international stage for their uproarious humour, pace and general verve.

These short plays mainly belong to the period 1888–1891. They belong, in other words, to the first years of Chekhov's maturity as a writer—assuming that one accepts the convenient and usual division of his career into two main sections: (*a*) a preparatory period of experiment and immaturity lasting about eight years (1880–7); and (*b*) a period of mature achievement, which covers the remaining years of his life (1888–1904).

So far as short stories are concerned, Chekhov produced a vast quantity of facetious pot-boilers during his early period—so many that his first eight years of literary activity account for writings about twice as bulky as those covered by the entire remaining seventeen years. But while Chekhov was writing furiously and frivolously during this early period, he was also quietly conducting (how deliberately and consciously it is hard to say) the literary experiments which were to enable him to blossom forth in 1888 as the most original Russian prose-writer of his age. Nor can the work of his early period by any means be dismissed as rubbish. Besides much worthless material, it includes minor masterpieces foreshadowing his characteristic later mode, apart from which many of his comic short stories are, within their own limited conventions, brilliantly executed.

How does Chekhov's dramatic work fit into the pattern of short-story writing just described? Not very neatly. In the early years he evidently took his plays more seriously than his short stories. His first two long plays (*Platonov*, ?1881; *Ivanov*, 1887–9) are both conceived seriously, as are the first two short plays to be found in the present volume: *On the High Road* (1885) and *Swan Song* (1887–8). These short plays are intensely tragic in conception, each being the study of a life broken by old age, unhappiness in love and alcoholism in different combinations. Thus Chekhov was trying his hand at high tragedy written for the stage at a time when most—though not all—of his short stories were frivolous and farcical. A closer look at the early short stories shows a fair number in which he attempts, exceptionally, to evoke a similar degree of despair: for example, the aptly-named *Sorrow* (*Горе*, 1885) and *Misery* (*Тоска*, 1886).

At this stage in his career Chekhov obviously felt an urge to write tragedy, but had little experience in setting about it. *On the High Road* and *Swan Song* are both 'tear-jerkers', the woes of Svetlovidov and Bortsov being somewhat over-emphasized. Who could imagine, looking at these early tragic works, that Chekhov would later become the supreme master of theatrical and fictional understatement? One might indeed feel sorry for a real-life Svetlovidov and Bortsov, but he must indeed have a heart of stone who, on meeting them in literary form, can contemplate their woes without being tempted to laugh.

Chekhov was later to discover that unhappiness could more effectively be expressed indirectly. 'Don't look sad . . .,' he wrote to his wife in 1901, advising her on the technique of playing Masha's part in his *Three Sisters*. 'People who have been unhappy for a long time, and grown used to it, don't get beyond whistling and are often wrapped up in their thoughts. So mind you look thoughtful fairly often on the stage.'[1] This is a far cry indeed from Svetlovidov's orgy of self-pity in *Swan Song*. But by the time when Chekhov wrote *Three Sisters* he had discovered that unhappiness could often be more effectively conveyed by what was left out than by what was put in, whereas this lesson still remained to be learned when he wrote *On the High Road* and *Swan Song*. To say this is not to deny that either of these two short plays is well made in its own way, or that both offer rich material for those 'fine old character actors' who have so long, rightly or wrongly, been the pride of the Russian stage.

One character in *On the High Road* calls for special comment: the tramp–highwayman Merik. This personage combines 'wind on the heath, brother' associations, brigandage and a well-camouflaged heart of gold. He also prefigures the kind of hero who was to be developed with enormous popular success by Gorky in the following decade: the romantic tramp–outlaw. Indeed, the play as a whole, with its ambience of picturesque deadbeats, seems to bear more resemblance to Gorky's

[1] See *The Oxford Chekhov*, vol. iii, p. 313.

Lower Depths than to a work by Chekhov. But Merik himself is not an entirely isolated phenomenon in Chekhov's work, having been preceded by at least one similar desperado: the horse-thief Osip in *Platonov*. A similar picaresque atmosphere is also evoked in the later short story *The Horse-stealers* (*Воры*, 1890). But Chekhov was ill at ease in this particular alley, which was suitable for exploration by coarser talents, and there is thus no cause to regret his eventual decision to neglect the study of colourful vagabonds. The episode does, however, help to underline once again a feature in his development which calls for repeated emphasis: his achievement of a mature and original mode was no matter of luck. It was the outcome of much ingenious and painstaking experiment with all sorts of material—some of it, in retrospect, startlingly 'un-Chekhovian'. Many were the barren avenues explored, many the cumbrous stones turned in the course of the search.

Occasional expressions of despair, recalling those in *Swan Song* and *On the High Road*, are to be found in other items contained in this volume. One of these is *Smoking is Bad for You*. As shown in Appendix ix, no fewer than six recensions of this short monologue have to be taken into account. The effect of the last two was to convert an originally farcical text into a mixture of farce and tragedy which well accords with the 'laughter through tears' formula so often invoked in discussions of Russian literature. Here the farcical conception belongs to the years 1886–9 and the infusions of despair to the 1890s (it is impossible to be precise in dating all the alterations to this text).

Tatyana Repin belongs to a genre all of its own (see Appendix v), but it too contains some incidental expressions of despair: those put into the mouths of the caretaker and of Father Ivan. ('We buried a local squire this afternoon, we've just had this wedding and we've a christening tomorrow morning. Where will it all end? What use is it to anyone? None, it's just pointless. . . . Oh, what a life, when you come to look at it. Why, I courted a girl myself once, married her and received a dowry, but that's all buried in the sands of time now.')

Such passages reflect an ultimate philosophic despair which was a recurring element in Chekhov's own mood—a part, but by no means the whole, of his outlook. He was also given to moods of buoyant optimism: witness his many happy predictions for the future of the human race. Contemplating the pessimism/optimism dichotomy which was such a fundamental element in Chekhov's make-up, his critics have been prone to fix exclusively on one or other pole. Hence the tendency to dismiss him either as a hopeless pessimist or an ebullient leader of his generation, a kind of fatuous activist. However, it was neither despair nor confident affirmation, but the tension between the two, which helped to make Chekhov into a great artist.

This having been said, we may turn from the two 'tear-jerker' plays and the oddity *Tatyana Repin* to the six completed farces in the present volume, which form Chekhov's chief contribution to the one-act drama. Here again a paradox is to be observed. Among his dramatic writings these farces (called 'vaudevilles' in Russsian) correspond in mood and style to the early comic short stories—whereas the four-act plays of Chekhov's maturity correspond to the serious short stories of his mature period. Chronologically speaking, however, the vaudevilles belong, as already mentioned, to the mature period—admittedly largely to the beginning of it. Hence the paradox of Chekhov's dramatic work: the early serious plays written when his prose fiction was largely comic—contrasting with these riotous vaudevilles produced at a period when he had turned his back entirely on facetiousness in prose fiction and had begun to produce the serious short stories of his maturity.

The farces (apart from the curious mixture in *Smoking is Bad for You*) are about as far removed as could be both from 'tear-jerkers' and serious tragedy. But they do contain some ingredients in common with Chekhov's mature full-length plays; in particular because so much of the humour depends on the ploy of mutual misunderstanding. The inability of human beings to communicate with each other is such an integral element in the thematic material of Chekhov's maturity that it scarcely calls for illustration. What is managed by innuendo, omission and half-statement in the full-length plays is put over in the vaudevilles with whole-hearted enthusiasm. Much of the fun in *The Bear* is based on this: Smirnov simply cannot understand that Mrs. Popov has no money in the house. Again, in *The Proposal*, the quarrelsome Natasha fails to notice that she is the recipient of a proposal of marriage which she is eager to accept. *A Tragic Role* also hinges on a misunderstanding: Tolkachov develops his tirade on the disadvantages of commuting from a holiday cottage with such ferocious eloquence that his speeches must be ranked among the finest harangues in Chekhov's work; yet the point of all this rhetoric is lost on Murashkin, a misunderstanding which leads to one of the best comic denouements in this volume.

The Anniversary likewise hinges on failure of communication, in which the two female characters are the main culprits: Mrs. Merchutkin cannot see that her affairs are no concern of the Bank, any more than Tatyana can appreciate that family gossip is of no interest to her husband and his cashier at a time when they are feverishly preparing for their fatuous anniversary celebration. *The Wedding* is less squarely based on major misunderstandings, though failure to communicate has its part to play here too: the guest of honour cannot understand that his naval reminiscences are of no interest to the other wedding guests, or that he has been invited to the party to play the role of General when he is only a Commander; whereas the fact that he is supposed to have accepted twenty-five roubles to lend dignity to the proceedings is understandably slow to penetrate the consciousness of this honest old salt.

Honesty and integrity are indeed in short supply among the characters in these farces, who are nearly all stupid, self-centred, grasping and affected in some combination. Here Aplombov's threat to 'make her [his bride] wish she'd never been born—*as I'm a man of honour*' (see p. 119 below) because his mother-in-law has defaulted on part of the dowry, provides a delicious example, particularly in the words here italicized. But though the farces contain many unprepossessing male characters, the palm of villainy goes to the women. *On the High Road* is built on Mary's cruel desertion of her infatuated husband on their wedding day: an act of heartlessness which is only compounded by her refusal to recognize him when she happens to meet him years later as a down-and-out in a wayside inn. *Swan Song* is free from any anti-feminist bias, unless one can count the refusal of Svetlovidov's inamorata to marry him on the ground that he was an actor. Mrs. Popov in *The Bear* is a fairly innocent character for a Chekhovian female, her main offence being the affectation of shutting herself up in her house in widow's weeds in order to make herself into a romantic figure. But her clash with Smirnov affords him the opportunity to indulge in some splendid tirades on women.

The Proposal brings out another undesirable feminine trait: an eagerness to get married at all costs, even if it means taking a husband as disagreeable as Lomov. *Tatyana Repin* is comparatively kind to the female sex, unless a willingness to create a disturbance by taking poison in church during someone else's wedding is to be entered in the charge sheet. But in *A Tragic Role* women are once more in the dock in the person of Tolkachov's wife, who does not appear on stage, but whose habits of loading her husband with errands, dragging him off to dances and amateur theatrical performances and keeping him awake in the middle of the night by practising duets with holiday-making tenors, all contribute elements to her husband's magnificent harangues. In *The Wedding* the men for once come off worse than the women, but the boot is back on the other foot again in *The Anniversary* and in *Smoking is Bad for You*, where Nyukhin's unseen wife is one of the best comic villainesses in the collection.

'All women, large or small,' says Smirnov in *The Bear*, 'are simpering, mincing, gossipy creatures. They're great haters. They're eyebrow-deep in lies. They're futile, they're trivial, they're cruel, they're outrageously illogical. And as for having anything upstairs [*taps his forehead*]—I'm sorry to be so blunt, but the very birds in the trees can run rings round your average blue-stocking.' This hardly represents Chekhov's considered view of the female sex, but it is worth noting the readiness with which he introduces such anti-feminist material in his writings—especially in the present volume. This is particularly important in view of the prevalent cosy stereotype of Chekhov as one who automatically 'loved people'. He was, in fact, very far from 'loving people', whether men or women, but if he had to choose between the two as objects of approval, he preferred men. 'Men and women are worth five copecks a pair,' he once wrote, 'but men are the fairer and the more intelligent.'[1] This remark was only a casual aside, thrown out in a private letter, but other evidence could be adduced to show that it more accurately sums up Chekhov's general attitude than any cliché about 'loving people'. Be this as it may, when it came to creating farcical situations women were fair game to Chekhov.

A less successful source of amusement in the farces is the constant harping on violent physical malaise, whereby at moments of crisis the various characters pretend to be having a heart attack, or to be about to faint, or to be subject to various violent aches and pains. This device might provide Chekhov with a means of ending a play when no other suggested itself; it is used this way in *The Proposal*. But on the whole the lover's embrace (the ending of *The Bear*) provides a better denouement.

The Bear and *The Proposal* are the only items in the volume without previous literary entanglements and involvements of some kind. In the case of *Smoking is Bad for You* there is the long sequence of early variants, and with *Tatyana Repin* there is Suvorin's four-act play with the same title to be taken into account. Each of the remaining six plays in the volume represents an adaptation of one or more earlier short

[1] Letter to A. S. Suvorin, 26 December 1888.

stories by Chekhov himself, as shown in the relevant appendixes. On the threshold of his mature period Chekhov is thus to be found reworking in dramatic form early farcical material from his immature period. But the plays are on a higher literary level than the stories from which they derive, as might be expected. Chekhov was, after all, a more experienced writer when he came to compose them. There is also the point that comic dialogue, at which Chekhov excelled, does not offer scope for certain defects which are often to be observed in his comic *narrative*, as found in the early short stories. This often takes the form of heavy badinage in the manner of Dickens at his worst. Compare, particularly, the dramatic fragment *The Night before the Trial*—unfortunately left unfinished, since it has the makings of a successful one-act farce—with the short story of the same name, which is marred by obtrusive facetiousness. As he grew older, Chekhov increasingly pruned such early extravagances in his work, even when this meant sacrificing some extremely funny lines.[1]

Compared with Chekhov's long plays from *The Seagull* onwards, with their original dramatic technique, the short farces offer comparatively conventional material. But they reveal Chekhov as a master of stage technique (including that of writing dialogue) almost from the beginning; though the defects of his earlier *Platonov* in this respect show that the skill had not come to him naturally, but was developed by trial and error.

The vaudevilles were a success on the stage from the beginning. They were also lucrative, which was why Chekhov once remarked: 'When I've written myself out I'm going to write vaudevilles and live on them. I think I could write about a hundred of them every year. Vaudeville subjects gush out of me like oil from the wells of Baku.'[2] The vaudevilles are superb examples of their own particular genre, and one may therefore regret that only some half a dozen in all trickled out instead of the hundreds predicted by Chekhov. Still, where drama was concerned, he was to prospect other gushers in new fields. These were to produce a yield of higher quality and susceptible of much greater refinement than the vaudevilles.

[1] See, for example, some of the material spoken by Mrs. Merchutkin in the early recensions of *The Anniversary*, below, p. 187.

[2] Letter to A. S. Suvorin, 23 December 1888.

ON THE HIGH ROAD

[*На большой дороге*]

A DRAMATIC STUDY IN ONE ACT

(1885)

CHARACTERS

TIKHON YEVSTIGNEYEV, keeper of an inn on the high road

SIMON BORTSOV, a ruined landowner

MARY, his wife

SAVVA, an old pilgrim

NAZAROVNA }
YEFIMOVNA } pious old women

FEDYA, a factory worker who is passing through

YEGOR MERIK, a tramp

KUZMA, a traveller

A POSTMAN

DENIS, a coachman

Pilgrims, drovers, travellers, etc

The action takes place in south Russia

TIKHON's inn. Right, the bar; shelves containing bottles. At the back of the stage, a door opening on the road, with a dirty red lantern hanging above it on the outside. The floor and benches by the walls are jammed with pilgrims and travellers. Many are sleeping in a sitting position for lack of room. It is late at night. As the curtain rises there is a clap of thunder, and a flash of lightning is seen through the open door.

SCENE I

[TIKHON *is behind the bar.* FEDYA *is sprawled on one of the benches, quietly playing an accordion. Near him sits* BORTSOV *in shabby summer clothes.* SAVVA, NAZAROVNA *and* YEFIMOVNA *have settled on the floor near the benches.*]

YEFIMOVNA [*to* NAZAROVNA]. Give the old man a shove, dear. He ain't long for this world, I reckon.

NAZAROVNA [*pulling the edge of* SAVVA*'s coat off his face*]. Good Christian sir, be you alive or be you dead?

SAVVA. What—me dead? I'm alive, I am. [*Raising himself on one elbow.*] Cover my legs, old thing. That's right. A bit more on the right side. That's right. Bless you.

NAZAROVNA [*covering* SAVVA*'s legs*]. Go to sleep, dearie.

SAVVA. Sleep? How can I? If only I had the patience to bear this agony, I could do without sleep. A sinner deserves no peace. What's that noise, my dear?

NAZAROVNA. Thunder, by God's grace. There's a howling wind and the rain's fair lashing down, beating on the roof and windows like it was peas from the pod. Hear it? The floodgates of heaven are opened. [*Thunder.*] Holy saints above us!

FEDYA. All that roaring, thundering, crashing—when will it ever end? Boom, boom—like the roar of the forest. Boom, boom! Like a dog's howling, the wind is. [*Hunches himself up.*] It's cold. My clothes are sopping wet and the door's wide open. [*Plays softly.*] My accordion's soaked, friends, there's no tunes left in it, or I'd let you have a proper basinful—fair take your breath away, it would! Great stuff! Quadrilles, polkas, say, or your Russian song and dance—I can do 'em all.

When I was waiter at the Grand Hotel in town, I never saved no money, but when it came to the accordion, I really knew my stuff. I play the guitar too.

VOICE FROM THE CORNER. A foolish man talks foolish talk.

FEDYA. Fool yourself! [*Pause.*]

NAZAROVNA [*to* SAVVA]. You should be lying in the warm now, dearie, and warming your poor leg. [*Pause.*] Old man, good Christian sir. [*Nudges* SAVVA.] Not a-dying, are you?

FEDYA. Better have a drop of vodka, grandpa. That'll put fire in your belly, it will, and ease your heart a bit. Here, have some.

NAZAROVNA. Less of the fancy talk, young man. The old fellow may be passing away, a-repenting of his sins, and you carry on like this, you and your accordion. Leave the music alone, you saucy creature.

FEDYA. Why are *you* bothering him then? He's proper poorly, and you—. Women's foolishness, this is. He can't speak harshly to you, him being a holy man. Pretty pleased with yourself, ain't you, because he listens to a ninny like you? Sleep on, granddad, don't you listen to her. Let her chatter away and to hell with her. A woman's tongue's like the devil's broom, it sweeps the wise and cunning out of the house. Let her go to hell. [*Throws up his arms.*] Hey, you're thin as a rake, grandpa. Not half you are—just like a dead skelington, with no life in you. Are you really dying?

SAVVA. Why should I die? God grant I don't die before my time. I'll have a bad turn, and then, God willing, I'll get up. The Blessed Virgin will see I don't die in foreign parts, I'll die at home.

FEDYA. Do you come from far?

SAVVA. I'm a Vologda man, a humble citizen of that town.

FEDYA. Where's Vologda?

TIKHON. T'other side of Moscow, in the Province of——

FEDYA. Phew, you have come a way, granddad. Done it all on foot, have you?

SAVVA. I have, lad. I've been to the shrine of St. Tikhon and I'm on my way to the Holy Mountains. Then, God willing, I'll go to Odessa. They say you can get a cheap passage from there to Jerusalem. Twenty-one roubles it costs, 'tis said.

FEDYA. Ever been in Moscow?

SAVVA. That I have! Half a dozen times.

FEDYA. Decent town, is it? [*Lights a cigarette.*] Worth-while?

SAVVA. There's plenty of holy shrines, boy. And where there's lots of shrines it's always worth-while.

BORTSOV [*goes up to the bar and speaks to* TIKHON]. Once again, for Christ's sake give me a drink.

FEDYA. The great thing is, a town should be clean. If it's dusty they should water the place, and if it's muddy they should clean it up. There should be tall houses, a theatre, police and cab-drivers, er—. I've lived in towns myself, I know what's what.

BORTSOV. Just a glass, just this little one. Chalk it up, I'll pay.

TIKHON. That's enough of that.

BORTSOV. Oh come on! Please!

TIKHON. You clear off!

BORTSOV. You don't understand me. Get this into your head, you clod, if there's one drop of grey matter in your country bumpkin's skull: it's not me that's asking—it's my guts, to put it in your yokel language. It's my illness that's asking, can't you see?

TIKHON. There's nothing *to* see. Go away.

BORTSOV. Look, if I don't get a drink now, see, if I don't satisfy my craving, I may do something awful. God knows what I may not do. You've seen lots of drunks since you started keeping a pub, you swine, surely you know what they're like by now! They're ill! Chain 'em up, beat 'em, stab 'em if you like—but let 'em have their vodka. Look, I'm asking you humbly! Please! I'm demeaning myself, my God, I am.

TIKHON. Let's have the money and you'll get your vodka.

BORTSOV. Where can I get money? It's all gone on drink, the whole damn lot. So what can I give you? All I have left is my overcoat, but that I can't give you because I have nothing on underneath. How about my cap. [*Takes off his cap and gives it to* TIKHON.]

TIKHON [*examining the cap*]. H'm. There's caps and caps. This one's like a sieve.

FEDYA [*laughs*]. It's a proper gentleman's cap, one to walk down the street in and take off to the young ladies. 'I say, what ho! How do you do?'

TIKHON [*gives the cap back to* BORTSOV]. I wouldn't take that filthy thing as a gift.

BORTSOV. Well, if you don't like it, give me a drink on tick. I'll be coming back from town and I'll bring you your five copecks, and may it choke you. Yes, choke you—may it stick in your throat! [*Coughs.*] I loathe you.

TIKHON [*banging his fist on the counter*]. Can't you leave me alone? Who do you think you are? Some kind of a crook? What brought you here?

BORTSOV. I want a drink. It's not me that wants it, it's my illness, see?

TIKHON. Don't aggravate me, or you'll find yourself getting some fresh air double quick!

BORTSOV. What can I do? [*Leaves the bar.*] What can I do? [*Reflects.*]

YEFIMOVNA. It's the devil tempting you. You take no notice, guv'nor. That's the Evil One a-whispering 'drink, drink, drink!' But you answer back: 'No, no, no!' Then he'll leave you be.

FEDYA. There's a ruddy great banging in your brain-pan, I'll bet, and your insides ain't feeling too good either. [*Roars with laughter.*] You're a funny one, sir. Lie down and sleep. No point in flapping round the place like a ruddy great scarecrow, this ain't no kitchen garden.

BORTSOV [*angrily*]. Shut up! No one asked your opinion, jackass.

FEDYA. You keep a civil tongue in your head! I know your sort—as if there weren't enough tramps like you on the high road! As for me being a jackass, when I fetch you one over the ear-hole you won't half roar—the storm won't be in it! Jackass yourself! Low scum! [*Pause.*] Bastard.

NAZAROVNA. The good old man may be praying and giving up his soul to God, while these blackguards are quarrelling and swearing. You should be ashamed of yourselves.

FEDYA. Stop snivelling, you old frump. Them as comes in taverns must put up with tavern ways.

BORTSOV. What am I to do? What can I do? How can I make him see? What more eloquence do I need? [*To* TIKHON.] This makes my blood run cold. Dear old Tikhon! [*Weeps.*] Tikhon, please.

SAVVA [*groans*]. There's a shooting pain in my leg, like a bullet of fire. Old woman——

YEFIMOVNA. What is it, dearie?

SAVVA. Who's that crying?

YEFIMOVNA. The squire.

SAVVA. Ask the squire to shed a tear for me, so I may die in Vologda. Tears can make a prayer work wonders.

BORTSOV. I'm not praying, granddad. And these aren't tears either—it's my heart's blood running out. [*Sits down by* SAVVA'*s feet.*] Heart's blood, I tell you. Anyway, this is a bit beyond you, a bit outside your dim horizons, old boy. You're a benighted lot.

SAVVA. But where are those who can see the light?

BORTSOV. There are some bright enough to understand, granddad.

SAVVA. There are for sure, son. The saints had the light, they understood all sorrows. No need to explain, they'd just understand. They can tell from the look in your eyes. And when they understand you, it's such a comfort, you might never have been troubled—it goes away like magic.

FEDYA. Have you really seen the saints?

SAVVA. I have, lad. It takes all sorts to make a world. There are sinners and there are the servants of the Lord.

BORTSOV. I can't make sense of this. [*Quickly gets up.*] Talk must make sense, but what sense is there in me at the moment? I just have an instinct, a thirst! [*Hurries to the bar.*] Tikhon, take my coat, do you hear? [*Makes to take his coat off.*] My overcoat——

TIKHON. What have you got underneath? [*Looks under* BORTSOV'*s overcoat.*] Your bare skin? Keep it, I won't have it. I won't take a sin upon my soul.

[MERIK *comes in.*]

SCENE II

[*The above and* MERIK.]

BORTSOV. Very well, I'll take the sin upon *my* soul. Agreed?

MERIK [*silently takes off his outer coat and stands there wearing a jerkin; he has an axe at his belt*]. There's them as feels the cold, but the bear and the homeless wanderer are always too hot. Fair sweating, I am. [*Puts the axe on the floor and takes off his jerkin.*] While you drag one foot out of the mud, you sweat a bucketful, and by the time one foot's free t'other's got stuck.

YEFIMOVNA. Quite right. Is it raining less, son?

MERIK [*with a glance at* YEFIMOVNA]. I don't bandy words with women. [*Pause.*]

BORTSOV [*to* TIKHON]. I'll take the sin upon myself, do you hear?

TIKHON. I don't want to hear, leave me alone.

MERIK. It's so dark—like as if the sky had a coat of tar. You can't see your nose in front of you, and the rain whips your face like it was a real old blizzard. [*Takes his clothes and axe in his arms.*]

FEDYA. It's a nice day for crime, ain't it? Even wild animals take cover, but you jokers are in your element.

MERIK. Who said that?

FEDYA. Take a look. Not blind, are you?

MERIK. I'll chalk that up to you. [*Goes up to* TIKHON.] Hallo there, you with the face! Or don't you know me?

TIKHON. Expect me to know all you drunken vagabonds? Reckon I'd need a dozen sets of eyes for that.

MERIK. Well, have a look. [*Pause.*]

TIKHON. But I do recognize you, I declare—I know you by your eyes. [*Shakes hands.*] Andrew Polikarpov, ain't it?

MERIK. It was, but now it's Yegor Merik, I'd say.

TIKHON. How come?

MERIK. Whatever label God gives me, that's my name. I've been Merik for a couple of months. [*Thunder.*] Ger! Thunder away, you don't scare me. [*Looks round him.*] No bloodhounds about?

TIKHON. Bloodhounds? Midges and gnats, more like! They're a soft lot. The bloodhounds will be snoring in their feather-beds, I reckon. [*Loudly.*] Watch your pockets, friends, and your clothes too, if you care about 'em. This is a bad man. He'll rob you.

MERIK. Let 'em look to their money if they have any, but clothes I won't touch, I've no use for 'em.

TIKHON. Where the devil are you heading for?

MERIK. The Kuban District.

TIKHON. Are you now!

FEDYA. The Kuban? Really? [*Sits up.*] That's wonderful country, lads, it's a land beyond your wildest dreams. Fine open country! There's no end of birds, they say, and game and all kind of beasts, and the grass, by God, grows all year round. The folks are real friendly like, and they've more land than they know what to do with. They say the government will let you have three hundred acres a head—or so a soldier was telling me t'other day. What luck, God damn me!

MERIK. Luck? Luck walks behind your back, you don't see it. Bite your elbow and you may glimpse it. It's all foolishness. [*Looks at the benches and people.*] You might be a chain-gang having a night off. Hallo there, down-and-outs!

YEFIMOVNA [*to* MERIK]. What vicious eyes you have! You have the devil inside you, boy. Don't look at us.

MERIK. What cheer, my beggarly chums!

YEFIMOVNA. Turn away. [*Pushes* SAVVA.] Savva dear, a bad man's looking at us. He'll hurt you, dearie. [*To* MERIK.] Turn away, you snake, I tell you.

SAVVA. He won't harm you, old woman, never fear. God won't let him.

MERIK. Hallo there, friends! [*Shrugs his shoulders.*] They don't speak. Not asleep, are you, you clumsy oafs? Why won't you speak?

YEFIMOVNA. Turn those great eyes away. And turn away from your satanic pride.

MERIK. Shut up, you old bitch! I wanted to give you a word of kindness and good cheer in your misery, there's no satanic pride in that. You're huddled up in the cold like a lot of flies—so I felt sorry for you and wanted to say a kind word and comfort you in your wretchedness, but you all turn your ugly mugs away. Ah well, a fat lot I care! [*Goes up to* FEDYA.] And where might you come from?

FEDYA. From these parts—the Khamonyev brickworks.

MERIK. Get up, will you?

FEDYA [*sitting up*]. What's that?

MERIK. Get up, man. Get right up, I want that place.

FEDYA. Oh, so this is *your* place, is it?

MERIK. Yes. You go and lie on the floor.

FEDYA. Out of my way, you tramp! You don't scare me.

MERIK. Quite a lively lad! Now you clear off and don't argue, you fool, or you'll be sorry.

TIKHON [*to* FEDYA]. Don't cross him, boy. Let him have his way and to hell with him.

FEDYA. What right have you got? Rolls his great fish eyes at me and thinks I'm scared. [*Gathers up his belongings in his arms and goes and makes a bed on the floor.*] Blast you! [*Lies down and pulls the clothing over his head.*]

MERIK [*makes his bed on the bench*]. Well, you don't know much about devils if you think I'm one. They're not like me. [*Lies down and puts the axe by his side.*] Lie there, little axe—let me cover up your handle.

TIKHON. Where did you get that axe?

MERIK. Stole it, I did, and now I seem stuck with the blasted thing—it seems a pity to throw it away, and I've nowhere to put it. It's like a wife you've got sick of, it is. [*Covers himself up.*] Devils aren't like me, lad.

FEDYA [*poking his head out from under his coat*]. Then what are they like?

MERIK. They're like steam or your breath. If you blow [*blows*]—they're like that. You can't see 'em.

VOICE [*from the corner*]. If you sit under a harrow, you'll see 'em.

MERIK. I've sat under one and seen none. That's an old wives' tale. Devils, pixies, ghosts—you can't see 'em. Our eyes ain't made to see everything. When I was a boy, I used to go into the forest at night, specially to see the pixies. I'd yell for a pixy at the top of my voice, and keep my eyes skinned. I'd see all sorts of funny things, but no pixies. Then I'd go to a churchyard of a night to look for ghosts—but that was all old wives' tales too. I saw various animals, but as for anything to scare me—nothing doing. The eye ain't made that way.

VOICE [*from the corner*]. Don't say that, it does happen. A peasant was gutting a pig in our village. He's cutting out the tripes when out pops one of 'em.

SAVVA [*sitting up*]. Don't talk of the devil, lads! It's a sin, dear boys.

MERIK. Aha, the old greybeard! Mr. Skelington! [*Laughs.*] No need for us to go to no churchyard, we've got our own ghosts crawling out of the woodwork to tell ùs where we get off! Sinful, he calls it. It's not for you to lay down the law—you and your stupid ideas! You're an ignorant, benighted lot. [*Lights his pipe.*] My father was a peasant and liked laying down the law too, at times. He steals a sack of apples from the priest one night and brings it to us with a sermon: 'Mind you kids don't scoff them apples before Harvest Festival, because that's a sin.' That's you all over—you won't speak the devil's name, but you can behave like devils incarnate. Take this old cow, for instance. [*Points to* YEFIMOVNA.] She saw me as the devil, but I'll bet she's sold her own soul to the devil half a dozen times over, on account of her woman's foolishness.

YEFIMOVNA. Ugh! May the power of the Cross be with us! [*Buries her face in her hands.*] Savva, dear.

TIKHON. Why try to scare us? Think you're clever, don't you? [*The door bangs in the wind.*] Christ, what a wind!

MERIK [*stretches*]. Oh, for a chance to show my strength! [*The door bangs in the wind.*] I'd like to match my strength with yon wind! It can't pull the door off, but for two pins I'd rip this whole inn up by the roots! [*Gets up and lies down again.*] Oh, I'm fed up.

NAZAROVNA. Say a prayer, you monster! Can't you stop fidgeting?

YEFIMOVNA. Leave him alone, blast him. He's looking at us again. [*To* MERIK.] Don't stare, bad man! Them eyes! He looks like Satan at his prayers!

SAVVA. Let him look, good women. Say a prayer and the evil eye can't touch you.

BORTSOV. Oh, I can't stand it, it's more than I can bear. [*Goes to the bar.*] Look here, Tikhon, for the last time, please—give me half a glass!

TIKHON [*shakes his head*]. Then where's the money?

BORTSOV. Oh God, I've told you once, haven't I? It's all gone on drink. Where do you think I can get any? It won't break you, will it, if you let me have a drop of vodka on tick? A glass of vodka hardly costs you anything, but it will save me from torture—torture, I tell you! I'm not just fussing, this is real suffering, can't you see?

TIKHON. Go and tell someone else about it, not me. Go and beg from these good Christian people—let them treat you if they like. Beggars get bread from me, nothing else.

BORTSOV. *You* take the poor creatures' money—I can't, I'm sorry. It's not my job to rob them, and I won't do it, see? [*Bangs his fist on the bar.*] I won't do it! [*Pause.*] Eh, wait a minute. [*Turns to the pilgrims.*] It's an idea, friends, I must say. Give me five copecks. My guts needs it, I'm ill.

FEDYA. So we're to stand you a drink, you dirty twister? How about a glass of water?

BORTSOV. The depths I've sunk to! Never mind, I don't want it—it was a joke.

MERIK. You'll get nothing from him, Squire, he's too stingy—we all know that. Wait a minute—I had five copecks somewhere about. We'll have a glass between us. [*Rummages in his pockets.*] Hell, it must be somewhere. I thought I heard a jingling in my pocket t'other day. No, I've nothing. Nothing doing, old man—it's just your bad luck. [*Pause.*]

BORTSOV. I must have a drink or I'll do something violent, I might kill myself. Oh God, what can I do? [*Looks out through the door.*] Shall I go away? Shall I go off into the blue—the black, rather?

MERIK. Why don't you preach at him, you pious old women! And you, Tikhon, why don't you kick him out? He hasn't paid for his night's lodging, has he? Get rid of him, chuck him out on his ear! People are that cruel nowadays, there's no gentleness and kindness in them. A lot of brutes, they are! If they see a drowning man, they shout: 'Go on then—drown. We've no time to watch—we've got our work to do.' As for throwing him a rope, not a chance! Ropes cost money.

SAVVA. Don't you be so hard on people, good sir.

MERIK. Shut up, you old horror! You're a savage lot! Monsters! Treacherous scum! [*To* TIKHON.] Come here and take my boots off! Jump to it!

TIKHON. I say, he has gone off the deep end! [*Laughs.*] Proper terror, aren't you?

MERIK. Come here, I tell you. And look sharp. [*Pause.*] Do you hear me? I'm not talking to thin air, am I? [*Gets up.*]

TIKHON. All right, all right—that'll do.

MERIK. I want you to pull my boots off, you vampire—the boots of a beggar and tramp.

TIKHON. Come on, come on—don't be so bad-tempered. Come and have a drink. Come on.

MERIK. What did I say I wanted, friends? A free vodka or my boots pulled off? Didn't I make myself clear, didn't I put it straight? [*To* TIKHON.] So you didn't catch what I said? I'll give you a minute—you'll catch it then all right.

[*This causes some stir among the pilgrims and travellers, who get up and look at* TIKHON *and* MERIK *in silent expectation.*]

TIKHON. What the hell brought you here! [*Comes out from the bar.*] Quite fancy yourself, don't you? Ah well, I'd better, I suppose. [*Takes off* MERIK'*s boots.*] You treacherous scum!

MERIK. That's right. And put them side by side. That's right. Now clear off.

TIKHON [*after taking off the boots, goes behind the bar*]. Think you're very clever, don't you? Any more of your tricks and you'll be out of this place on your neck, I'm telling you! [*To* BORTSOV, *who comes up to him.*] You again?

BORTSOV. Look, I've got something in gold I might let you have. All right then, if you like, I'll give you——

TIKHON. Why are you shaking? Talk sense.

BORTSOV. It's a rotten, low thing to do, but there's nothing for it. It's a dirty trick I have in mind, but then my mind's none too sound—any court would acquit me. Take it, but on this condition: return it to me when I get back from town. I give it before witnesses. Now witness this, all of you. [*Gets a gold locket from an inside pocket.*] Here. I ought to take the picture out, but I've nowhere to put it, I'm wet through. Come on, take the lot—the picture too! But one thing: don't, er, put your fingers on the face. Please. I was rude to you, my dear fellow, and silly, but forgive me and don't finger it. I don't want your eyes looking at the face. [*Gives* TIKHON *the locket.*]

TIKHON [*examines the locket*]. A stolen watch. All right then, have your drink. [*Pours some vodka.*] Put that inside you!

BORTSOV. Only don't you, er, touch—. [*Drinks slowly with shuddering pauses.*]

TIKHON [*opens the locket*]. I see—a lady. Where did you pick up a bit of stuff like that?

MERIK. Show it here. [*Gets up and goes to the bar.*] Give me a look.

TIKHON [*pushes his hand away*]. What do you think you're doing? Look at it while I hold it.

FEDYA [*gets up and goes to* TIKHON]. Let's have a look.

[*The pilgrims and travellers go up to the bar from various directions and form a group.*]

MERIK [*firmly clasps in both his hands* TIKHON'*s hand which holds the locket, and looks at the portrait in silence; pause*]. Handsome little devil! A lady and all——

FEDYA. Not half—. Look at those cheeks and eyes! Move your fingers out a bit, I can't see. Hair down to the waist. It's just like she was alive and going to speak. [*Pause.*]

MERIK. That's the road to ruin for a weak man. Once saddled with a woman like that [*gives a gesture of despair*]—and you're dished!

KUZMA [*off-stage*]. Whoa! Stop, you stupid creature!

[*Enter* KUZMA.]

SCENE III

[*The above and* KUZMA.]

KUZMA [*comes in*]. 'The inn, a place where one and all
Who use this road are bound to call.'

Yes, you may drive past your dear old father in broad daylight without seeing him, but an inn—that you can spot a hundred miles off in the dark. Out of my way, all true believers! You there! [*Bangs a coin on the bar.*] A glass of real madeira. And make it snappy!

FEDYA. Hey, you seem in one hell of a hurry!

TIKHON. Don't wave your arms, you'll knock something over.

KUZMA. What are arms for but to wave? What are you so scared of, you ruddy wilting lilies? A spot of rain, poor dears? [*Drinks.*]

YEFIMOVNA. It's enough to scare anyone, good sir, to be caught on the road on a night like this. We're well off these days, praise the

Lord, with plenty of villages and farms on the way, so you've somewhere to shelter from the weather. But in the old days things was past praying for. Seventy miles you might travel, and not see a single twig, let alone a village or farm. You had to sleep rough.

KUZMA. How long have you been knocking round, old girl?

YEFIMOVNA. I'm past seventy, sir.

KUZMA. Seventy! You'll soon be in your second childhood then. [*Looks at* BORTSOV.] What queer fish is this? [*Stares at* BORTSOV.] A gentleman, eh?

[BORTSOV *recognizes* KUZMA, *goes into a corner, embarrassed, and sits on a bench.*]

KUZMA. Mr. Bortsov, is it you—yes or no? What are you doing in this dump? This is no place for you, surely.

BORTSOV. Hold your tongue.

MERIK [*to* KUZMA]. Who is he?

KUZMA. A most unhappy man. [*Walks nervously up and down by the bar.*] Eh? In a low dive—I ask you! In rags! Drunk! Oh, this has given me quite a turn, it has that! [*To* MERIK, *in a half whisper.*] This is the guv'nor, our landlord—Mr. Simon Bortsov, Esquire. See what a state he's in? He looks like nothing on earth! That's what the drink does for you. Fill me up, will you? [*Drinks.*] I come from his village, Bortsovka—you may have heard of it, it's about a hundred and fifty miles from here, in the Yergov District. We were his father's serfs. It's a rotten shame.

MERIK. Was he rich?

KUZMA. Oh yes, he was a big man.

MERIK. Squandered his father's money, did he?

KUZMA. No, it was sheer bad luck, man. He was a grand gentleman—rich, never the worse for drink. [*To* TIKHON.] You must have seen him yourself at times, I reckon, driving past the inn on his way to town. He had proper squire's horses—real nippy, they were—and a carriage with springs, all high-class stuff. He kept five troikas, man. I remember him crossing by the Mikishkin ferry hereabouts five years ago, and tossing them a rouble instead of five copecks. 'No time to wait for change,' says he. Not bad, eh?

MERIK. He must have gone out of his mind then.

KUZMA. He still seems to have his wits about him. It all came from being so feeble. And spoilt. The great thing is, it was all woman's work, lads. The poor man falls in love with a girl in town, fancies she's the loveliest thing in creation. She ain't no fairy princess, but he loves her like she was, see? She was a young lady, though, not a loose woman or anything like that, but a giddy little thing. Oh, she was proper flighty, she was—screwing her eyes up and laughing and all that. No sense, she had. The gentry like that kind—think they're real clever, when none of us peasants would give 'em house room. Well, the squire takes a fancy to her, and it's all up with him. He starts carrying on with her, and one thing leads to another and so on, with them going boating all night and playing pianos and that.

BORTSOV. Don't tell them, Kuzma. What's the point? What business is it of theirs, how I've lived?

KUZMA. Sorry, sir, I've only told them a bit of it. I've had my say, and that's all they'll hear. I told them that bit because I was upset like, oh, very upset I was. Come on, fill it up. [*Drinks.*]

MERIK [*in a half-whisper*]. And did she love him?

KUZMA [*in a half-whisper which gradually turns into ordinary speech*]. Well, what do you think? It's not as if the squire was just anyone! What—her not fall in love, and him with his couple of thousand acres and money to burn! And he was that respectable, too—dignified and well-behaved, like—and well in with the nobs. Like this. [*Takes* MERIK'*s hand.*] 'Oh, I say, what ho! Cheerio! Oh, do come in!' Well, one night I'm walking through the squire's garden—and, brother, what a garden! Miles and miles of it. I'm walking along quietly, keeping my eyes open, and there they are sitting on a bench [*makes the sound of a kiss*], a-kissin' of each other. He kisses her once and she kisses him twice, the little bitch. He takes her little white hand, and she blushes and snuggles up to him, drat 'er! 'I love you, Simon,' says she. And Simon goes about like a lunatic, boasting how happy he is, being a bit weak in the head, like. Gives a rouble here and two roubles there. Gives me money to buy a horse. Lets everyone off their debts, he's so pleased with hisself.

BORTSOV. Oh—why go into all this? These people have no feelings. It's painful, can't you see?

KUZMA. I've not said much, sir. They keep on at me. Why not tell them a bit? All right then, I won't if it makes you angry. I won't. I don't give a damn for 'em.

[*The noise of mail-coach bells is heard.*]

FEDYA. Don't yell, tell it quietly.

KUZMA. I *am* telling it quietly. Can I help it if he wants me to stop? There's no more to tell, anyway. They got married and that was that. Pour a glass for good old Kuzma! [*Drinks.*] I don't like drunkenness. Just as the ladies and gentlemen are going to sit down to the wedding breakfast, she ups and rushes off in a carriage. [*In a whisper.*] Dashes off to town to a lawyer-fellow who's her lover. How do you like that, eh? She certainly picked her moment! Killing's too good for her, I'd say.

MERIK [*pensively*]. I see. And what happened next?

KUZMA. He goes clean off his rocker. He went on the booze, as you see, and he's never looked back since, they say. First it was little ones, now he's got to pink elephants. Still loves her, he does. Look how he loves her. He's walking all the way to town now just to have a peep at her, I reckon. Then he'll come back.

[*The mail-coach drives up to the inn. The* POSTMAN *comes in and has a drink.*]

TIKHON. The mail's late today.

[*The* POSTMAN *pays and goes out without speaking. The coach drives off with bells jingling.*]

VOICE [*from the corner*]. Just the weather for a mail robbery—it would be dead easy.

MERIK. I've been around for thirty-five years and never robbed the mail yet. [*Pause.*] Now it's gone and I'm too late. Too late I am.

KUZMA. Feel like a sniff at Siberia, do you?

MERIK. Not everyone gets caught. Anyway, I wouldn't mind. [*Abruptly.*] And what happened next?

KUZMA. You mean to this poor fellow?

MERIK. Why, who else?

KUZMA. Well, friends, the next thing that helped to ruin him was his brother-in-law, his sister's husband. He takes it into his head to back this brother-in-law, a bank loan to the tune of thirty thousand. The brother-in-law's a regular shark—the swine knows which side his bread's buttered on, of course, and he don't bat an eyelid. Borrows the money, but don't feel obliged to pay it back. So the master pays up all thirty thousand. [*Sighs.*] A fool and his money are soon parted. His wife has children by this lawyer-man, Brother-in-law buys an estate near Poltava, and our friend crawls round like an idiot from one low dive to another a-moaning and a-groaning to the likes of us: 'I've lost my faith, friends. I don't trust no one now!' Sheer weakness, I call it. We've all got our troubles nagging at us, but that don't mean we have to drown 'em in drink, do it? Take our village elder, now. His wife entertains the schoolmaster in broad daylight and spends her husband's money on booze, while her old man goes round with a grin on his face. He has got a bit thin, though.

TIKHON [*sighs*]. It depends how much strength God gives you.

KUZMA. Some are stronger than others, I grant you. Well, how much? [*Pays.*] Take my hard-earned cash. Good-bye, boys. Good night and pleasant dreams. It's time I was off. I'm fetching a midwife from hospital for the missus. Poor woman must be tired of waiting and wet through. [*Runs out.*]

TIKHON [*after a pause*]. Hey, you—what's your name? Come and have a drink, poor fellow. [*Pours one out.*]

BORTSOV [*comes up hesitantly to the bar and drinks*]. So I owe you for two glasses.

TIKHON. Owe me? Rubbish! Drink up and forget it. Drown your sorrows.

FEDYA. Have one on me too, Squire! Ah well! [*Throws a five-copeck piece on the bar.*] You die if you drink and you die if you don't. Life's all right without vodka—still, vodka does put a bit of life in you, by golly! Grief ain't grief when you've a drink inside you. Swill it down!

BORTSOV. Phew, hot stuff that!

MERIK. Give it here! [*Takes the locket from* TIKHON *and examines the portrait.*] I see. Ran away on her wedding day. Quite a girl!

VOICE [*from the corner*]. Give him another, Tikhon. Let him have one on me.

MERIK [*bangs the locket on the floor violently*]. Blast her!

[*Goes quickly to his place and lies down, face to the wall. General consternation.*]

BORTSOV. What's that? What's going on? [*Picks up the locket.*] How dare you, you swine? What right have you to do that? [*Tearfully.*] Want me to break your neck, you clumsy lout?

TIKHON. Don't be angry, Squire. It ain't made of glass, it's not broken. Have another, and sleep it off. [*Pours another glass.*] I'm tired of all your talk, and it's long past closing-time. [*Goes and shuts the outer door.*]

BORTSOV [*drinks*]. How dare he? The idiot! [*To* MERIK.] Know what you are? An idiot! A jackass!

SAVVA. Won't you please curb your tongues? Why make such a row, good friends? Let folk sleep.

TIKHON. Lie down, lie down. That will do. [*Goes behind the bar and locks the till.*] It's bedtime.

FEDYA. It is that. [*Lies down.*] Happy dreams, boys.

MERIK [*stands up and spreads a fur coat on the bench*]. Come and lie down, Squire.

TIKHON. But where will you sleep?

MERIK. Anywhere, I don't mind the floor. [*Spreads his coat on the floor.*] I don't care. [*Puts his axe by his side.*] The floor's agony to him, being used to silk and soft bedding and such.

TIKHON [*to* BORTSOV]. Lie down, sir. Don't look at that there picture any more. [*Puts out the candle.*] Forget her!

BORTSOV [*staggering*]. Where shall I lie?

TIKHON. In the tramp's place. He wants you to have it, didn't you hear?

BORTSOV [*goes to the bench*]. I, er—I'm drunk. What's this? I'm to lie here, am I?

TIKHON. Yes, here, don't be afraid—lie down. [*Stretches out on the bar.*]

BORTSOV [*lies down*]. I'm drunk. My head's going round. [*Opens the locket.*] Have you a bit of candle? [*Pause.*] You're a funny girl, Mary, looking out of the frame and laughing at me. [*Laughs.*] I'm drunk—you shouldn't laugh at a drunkard. But don't you take no notice, as the man says in the play. You love the poor old soak.

FEDYA. How the wind howls—scares you, don't it?

BORTSOV [*laughs*]. You are a funny girl! Why twist and turn like that? I can't catch you.

MERIK. He's raving. He's been looking at that portrait too long. [*Laughs.*] What a business! Brainy gents have invented machines and medicines galore, but what about a cure for the female sex? No one ain't had the brains to invent that. They try to cure every illness, but one thing's never even crossed their minds: there's more men comes a cropper over a bit of skirt than from any illness. They're sly and cruel, women are, they're out for what they can get. And they ain't got no sense. The old woman torments her son's wife, and the girl herself never stops trying to do her husband down. And so it goes on.

TIKHON. Women have led him such a dance, he's still dizzy.

MERIK. I ain't the only one. Men have been suffering since time began. And why is it that women and the devil always go together in fairy-tales and songs? That ain't no accident, believe you me. It's more than half true, it is. [*Pause.*] There's the squire making a fool of himself, but what about me turning tramp and leaving my father and mother? That wasn't too clever either.

FEDYA. Women's doing, eh?

MERIK. Same as the squire here. I went round like I was mad or bewitched—boasted how happy I was. It was like being on fire day and night, but when the time came my eyes was opened. It weren't love, it were all a fraud.

FEDYA. What did you do to her?

MERIK. None of your business. [*Pause.*] Think I killed her, eh? My arms ain't that long. Killed her? Felt sorry for her, more like! 'Live and be happy,' I tells her. 'Only don't let me set eyes on you, and may I forget you, you treacherous bitch.'

[*A knock on the door.*]

TIKHON. Who the devil's that? Who's there? [*A knock.*] Who's knocking? [*Gets up and goes to the door.*] Who's knocking? Move on, we're locked up.

DENIS [*off-stage, on the other side of the door*]. Let us in, Tikhon, for goodness' sake. A spring's gone in the carriage. Help me, please. I only want a bit of rope to tie it up and I'll get by somehow.

TIKHON. Who is it then?

DENIS [*off-stage*]. A lady from town on her way to Varsonofyevo, with only three miles to go. Help us, please!

TIKHON. Go and tell your lady—for ten roubles she'll have her rope and we'll mend the spring.

DENIS [*off-stage*]. Are you mad? Ten roubles! You must be crazy! Glad to see folks in trouble, are you?

TIKHON. Have it your own way. You can take it or leave it.

DENIS [*off-stage*]. Oh, all right then, wait. [*Pause.*] The lady says yes.

TIKHON. Come in then. [*Opens the door and lets* DENIS *in.*]

SCENE IV

[*The above and* DENIS.]

DENIS. Hallo, friends. Well, let's have the rope. Hurry up. Who'll come and help, boys? We'll make it worth your while.

TIKHON. Never mind that. Let 'em snore, we'll manage between us.

DENIS. Phew, I'm dead beat, what with the cold and mud and being soaked to the skin. Another thing—have you a room for the lady to warm herself? The carriage is down on one side and she can't sit there.

TIKHON. What—a room is it now? She can get warm in here if she's cold, we'll find space for her. [*Goes up to* BORTSOV *and clears the space next to him.*] Get up there. You can doss down on the floor for an hour while the lady's getting warm. [*To* BORTSOV.] Get up a moment, please, sir. And sit for a bit. [BORTSOV *sits up.*] There's a place for you.

[DENIS *goes out.*]

FEDYA. So we've a visitor now, blast her! Now we won't get a wink till daybreak.

TIKHON. I'm sorry I didn't ask fifteen roubles, she'd have paid it. [*Stands before the door expectantly.*] Be on your best behaviour, all of you, and no bad language.

[*Enter* MARY *followed by* DENIS.]

SCENE V

[*The above,* MARY *and* DENIS.]

TIKHON [*bows*]. Come in, lady. This is a humble place, fit for peasants and black beetles. But don't you be put out.

MARY. I can't see a thing here. Where do I go?

TIKHON. This way, lady. [*Takes her to the seat next to* BORTSOV.] This way, please. [*Blows on the seat.*] I've no separate room, sorry, but don't worry, lady—they're nice, quiet folk.

MARY [*sits down next to* BORTSOV]. I say, isn't it stuffy! Can't we at least have the door open?

TIKHON. Very well, lady. [*Runs and opens the door wide.*]

MERIK. Folks are freezing and they must have the door wide open. [*Gets up and slams it.*] Who does she think she is! [*Lies down.*]

TIKHON. I'm sorry, lady, this is a kind of village idiot, like. But don't be frightened, he won't do no harm. Only I can't manage this for ten roubles, missus—sorry. I can do it for fifteen if you like.

MARY. Very well, but be quick.

TIKHON. This minute. It'll only take a jiffy. [*Brings out some ropes from under the bar.*] This very instant. [*Pause.*]

BORTSOV [*stares at* MARY]. Mary. Mary——

MARY [*looking at* BORTSOV]. What is it now?

BORTSOV. Mary—is it you? Where have you come from?

[MARY *recognizes* BORTSOV, *shrieks and jumps away into the middle of the room.*]

BORTSOV [*follows her*]. Mary, it's me. Me. [*Roars with laughter.*] My wife! Mary! But where am I? Let's have some light, you there!

MARY. Leave me alone. You must be lying, it isn't you, it can't be! [*Buries her face in her hands.*] This is some silly trick.

BORTSOV. Her voice, the way she moves! Mary, it's me. I won't be, er, drunk in a moment. My head's going round. My God! Just a moment, nothing makes any sense. [*Shouts.*] My wife! [*Falls at her feet and sobs. A group gathers round the couple.*]

MARY. Go away from me. [*To* DENIS.] We're leaving, Denis, I can't stay here.

MERIK [*jumps up and stares at her face*]. The portrait! [*Clutches her arm.*] It's her! Hey, all of you—it's the squire's wife!

MARY. Leave me alone, you lout. [*Tries to tear her hand away.*] Don't just stand there, Denis. [DENIS *and* TIKHON *run up to her and seize* MERIK *under the arms.*] This is a den of thieves. You let go my arm. I'm not scared. Go away!

MERIK. Wait a moment and I'll let you go. I want a word with you, to make you understand, so wait a minute. [*Turns to* TIKHON *and* DENIS.] Go away, and take your filthy hands off of me! I'm not letting her go till I've had my say. Just wait a minute. [*Bangs his forehead with his fist.*] God, I'm so stupid—can't think what I want to say.

MARY [*pulls her arm away*]. Go away, you're all drunk. We're leaving, Denis. [*Makes for the door.*]

MERIK [*blocks her way*]. You might spare him a glance. At least say one kind word to him, in God's name!

MARY. Take this maniac away!

MERIK. Then the curse of hell be on you, blast you!

[*Swings his axe. A frightful commotion. Everyone jumps up noisily. Shouts of horror.* SAVVA *stands between* MERIK *and* MARY. DENIS *shoves* MERIK *violently to one side and carries his mistress out of the inn. Then all stand rooted to the spot. A long pause.*]

BORTSOV [*clutches the air with his hands*]. Mary. Where are you, Mary?

NAZAROVNA. My God, my God, you've made my heart bleed, you murderers. What a dreadful night!

MERIK [*dropping the hand which holds the axe*]. Did I kill her or not?

TIKHON. You're in the clear, praise the Lord.

MERIK. I didn't kill her, so—. [*Staggers to his place.*] So I'm not to die through a stolen axe. [*Falls on his coat and sobs.*] I'm so fed up, so damn miserable—aren't you sorry for me, all of you?

CURTAIN

SWAN SONG
(CALCHAS)

[*Лебединая песня* (*Калхас*)]

A DRAMATIC STUDY IN ONE ACT

(1887–1888)

CHARACTERS

VASILY SVETLOVIDOV, a comic actor, aged 68

NIKITA, a prompter, an old man

The action takes place at night on the stage of a provincial theatre after a performance

The empty stage of a second-class provincial theatre. Right, a row of roughly-made unpainted doors leading to the dressing-rooms. The left and the back of the stage are cluttered up with litter and rubbish. There is an overturned stool in mid-stage. Night. Dark.

SCENE I

[SVETLOVIDOV, *in the stage costume of Calchas, comes out of a dressing-room carrying a candle, and roars with laughter.*]

SVETLOVIDOV. This is the limit, it really is too much—falling asleep in my dressing-room! The play ended hours ago, the audience went home—and there's me snoring away without a care in the world. Oh, you silly old man—you are a bad lad, old boy. Got so pickled, you dozed off in your chair! Very clever! Congratulations, old boy. [*Shouts.*] George! George, curse you! Peter! They're asleep, damn them, blast them and may they rot in hell! George! [*Picks up the stool, sits on it and puts a candle on the floor.*] And answer came there none—apart from the echo, that is. I tipped George and Peter three roubles each today for looking after me, and by now they must be sunk without trace. They've left. And the bastards must have locked up the theatre. [*Twists his head about.*] Ugh, I'm drunk. God, the booze I knocked back in honour of my benefit night! I feel as if I'd been kippered, my mouth's like the bottom of a parrot's cage. Disgusting! [*Pause.*] And stupid! The old codger gets drunk, but what has he to celebrate? He hasn't the foggiest! God, I've got back-ache, the old head-piece is splitting, I'm shivering all over, and I have this dark, cold feeling, as if I was in a cellar. If you won't spare your health, you might at least remember you're too old for this caper, you silly old so-and-so. [*Pause.*] Old age—whether you try to wriggle out of it or make the best of it or just act the fool, the fact is your life's over. Sixty-eight years down the drain, damn it! Gone with the wind! The cup's drained, there's just a bit left at the bottom: the dregs. That's the way of it, that's how it is, old man. Like it or not, it's time you rehearsed for the part you play in your coffin. Good old death's only just round the corner. [*Looks in front of him.*] I say, I've been on the stage for forty-five years, but this must be the first time I've seen a theatre in the middle of the night, the very first time. You know, it's weird, damn it. [*Goes up to the footlights.*] Can't see a thing. Well, I can just make out the prompter's box and that other box over there—the one with the letter on it—and that music-stand. The rest is darkness, a bottomless black pit like a tomb: the haunt of Death itself. Brrr! It's cold, there's a piercing draught from the auditorium. Just the place to call up spirits! It's eerie, blast it, it sends shudders down my spine. [*Shouts.*] George! Peter! Hell, where are you? But why do I talk of hell, God help me? Oh, why can't you stop drinking and using bad language, for God's sake, seeing you're old and it's time you were pushing up the daisies? At sixty-eight people go to church, they get ready to die, but you—. Oh Lord, you and your bad language and your drunken gargoyle's face and this damfool costume! What a sight! I'll go and change quickly. It's all so eerie. Why, if I stayed here all night, I'd die of fright at this rate. [*Makes for his dressing-room. At that moment* NIKITA, *wearing a white dressing-gown, appears from the furthest dressing-room at the back of the stage.*]

SCENE II

SVETLOVIDOV [*seeing* NIKITA, *gives a terrified shriek and staggers back*]. Who are you? What are you after? Who do you want? [*Stamps.*] Who are you?

NIKITA. It's me, sir.

SVETLOVIDOV. Who's me?

NIKITA [*slowly approaching him*]. It's me. Nikita, the prompter. It's me, Mr. Svetlovidov, sir.

SVETLOVIDOV [*collapses helplessly on the stool, breathes hard and shudders all over*]. God, who is it? Is it you—you, Nikita? W-w-what are you doing here?

NIKITA. I always spend the night here in the dressing-rooms, sir, only please don't tell the manager. I've nowhere else to sleep, and that's God's truth.

SVETLOVIDOV. So it's you, Nikita. Hell, I had sixteen curtain-calls, three bunches of flowers and a lot of other things—they were all quite carried away, but no one bothered to wake the old soak up and take him home. I'm old, Nikita. Sixty-eight, I am. I'm ill. I feel faint and weary. [*Leans over the prompter's hand and weeps.*] Don't leave me, Nikita. I'm old and weak and I've got to die. I'm frightened, so terribly frightened.

NIKITA [*gently and respectfully*]. It's time you went home, Mr. Svetlovidov, sir.

SVETLOVIDOV. I won't go. I haven't got any home—haven't got one, I tell you.

NIKITA. Goodness me, the gentleman's forgotten where he lives!

SVETLOVIDOV. I don't want to go to that place, I tell you. I'm on my own there, I haven't anyone, Nikita—no old woman, no children, neither kith nor kin. I'm as lonely as the wind on the heath. There will be no one to remember me when I die. I'm frightened all alone. There's no one to comfort me, to make a fuss of me and put me to bed when I'm drunk. Where do I belong? Who needs me? Who loves me? No one loves me, Nikita.

NIKITA [*through tears*]. Your audiences love you, Mr. Svetlovidov.

SVETLOVIDOV. The audience has left and gone to bed, and it's forgotten the old clown. No, nobody needs me, no one loves me. I've neither wife nor children.

NIKITA. Then you've nothing to worry about.

SVETLOVIDOV. I'm a man, aren't I? I'm alive. I have blood, not water, flowing in my veins. I'm a gentleman, Nikita, I'm well connected, and I was in the army before landing up in this dump. I was a gunner. And a fine, dashing, gallant, high-spirited young officer I was. Ye gods, what's happened to all that? And what an actor I became, eh, Nikita? [*Hoists himself up and leans on the prompter's arm.*] What's become of it all, where have those days gone to? God, I just looked into this black pit and it all came back to me! It's swallowed forty-five years of my life, this pit has—and what a life! Looking into it now, I see everything down to the last detail as plain as I see your face. To be gay, young, confident, fiery! And the love of women! Women, Nikita!

NIKITA. It's time you were in bed and asleep, Mr. Svetlovidov, sir.

SVETLOVIDOV. When I was a young actor and just getting into my stride, there was a girl who loved me for my acting, I remember. She was elegant, graceful as a young poplar, innocent, unspoilt. And she seemed all ablaze like the sun on a May morning. Those blue eyes, that magic smile could banish the darkest night. The ocean waves dash themselves against the cliffs, but against the waves of her hair the very cliffs, icebergs and snow avalanches might dash themselves to no avail. I remember standing before her as I stand before you now. She was looking lovelier than ever, and she gave me a look I shan't forget even in the grave. There was a kind of soft, deep, velvety caress about it, and all the dazzle of youth. Drunk with joy, I fall on my knees and beg her to make me happy. [*Continues in a broken voice.*] And she—she tells me to leave the stage. Leave the stage, see? She could love an actor, but be an actor's wife? Never! I remember how I acted that same night. It was a vulgar, slapstick part, and I could feel the scales fall from my eyes as I played it. I saw then that there's no such thing as 'sacred art', that the whole thing's just a phoney racket—saw myself a slave, a toy for people's idle moments, a buffoon, a man of straw. It was then I took the public's measure. Applause, bouquets, wild enthusiasm—I've never believed in 'em since. Yes, Nikita, these people cheer me, they pay a rouble for my photograph, but to them I'm a stranger, I'm just so much dirt—an old whore, practically! They scrape up acquaintance with me to make themselves feel important, but not one would sink to letting me marry his sister or daughter. I don't trust 'em. [*Sinks on to the stool.*] Don't trust 'em.

NIKITA. You look like nothing on earth, Mr. Svetlovidov, sir—you've even scared me. Have a heart and let me take you home, sir.

SVETLOVIDOV. Then my eyes were opened, but the vision cost me dear, Nikita. After that affair—with the girl—I began drifting aimlessly, living from hand to mouth with no thought for the morrow. I took cheap, slapstick parts and hammed them. I was a corrupting influence. But I'd been a true artist, you know, I was really good! I buried my talent, cheapened it. I spoke in an affected voice, I lost my dignity as a human being. This black pit swallowed me up and gobbled me down. I never felt like this before. But when I woke up tonight, I looked back—and I've sixty-eight years behind me. I've only just seen what old age means. The show is over. [*Sobs.*] You can ring down the curtain!

NIKITA. Mr. Svetlovidov, sir. I say, really, old man. Do calm yourself, sir. Oh goodness me! [*Shouts.*] Peter! George!

SVETLOVIDOV. And what flair, what power, what a delivery! The wealth of feeling and grace, the gamut of emotions here in this breast [*beats his breast*]—you simply can't imagine! I feel like choking. Listen, old man—wait, let me get my breath. Here's something from *Boris Godunov*:

> 'Ivan the Terrible pronounced me son.
> And from the grave his spirit named me Dmitry;
> He stirred the peoples to revolt for me
> And destined Godunov to die my victim.
> I am Tsarevich. But enough! 'Tis shame
> To cringe before a haughty Polish beauty!'

Not bad, eh? [*Eagerly.*] Wait, here's something from *King Lear*. There's a black sky, see, and rain, with thunder growling and lightning whipping across the heavens, and he says:

> 'Blow, winds, and crack your cheeks! rage! blow!
> You cataracts and hurricanoes, spout
> Till you have drench'd our steeples, drown'd the cocks!
> You sulphurous and thought-executing fires,
> Vaunt-couriers to oak-cleaving thunderbolts,
> Singe my white head! And thou, all-shaking thunder,
> Strike flat the thick rotundity o' the world!
> Crack nature's moulds, all germens spill at once
> That make ingrateful man!'

[*Impatiently.*] Quick, the Fool's cue! [*Stamps.*] The Fool's cue and quick about it! I'm in a hurry.

NIKITA [*playing the part of the Fool*]. 'O nuncle, court holy-water in a dry house is better than this rain-water out o' door. Good nuncle, in, and ask thy daughters' blessing; here's a night pities neither wise man nor fool.'

SVETLOVIDOV. 'Rumble thy bellyfull! Spit, fire! spout, rain!
Nor rain, wind, thunder, fire, are my daughters:
I tax not you, you elements, with unkindness;
I never gave you kingdom, called you children.'

What power, what genius, what an artist! Now for something else—something else to bring back old times. Let's take something [*gives a peal of happy laughter*] from *Hamlet*. All right—I commence! But what shall I do? Ah, I know. [*Playing* Hamlet.] 'O! the recorders: let me see one.' [*To* NIKITA.] 'Why do you go about as if you would drive me into a toil?'

NIKITA. 'O! my lord, if my duty be too bold, my love is too unmannerly.'

SVETLOVIDOV. 'I do not well understand that. Will you play upon this pipe?'

NIKITA. 'My Lord, I cannot.'

SVETLOVIDOV. 'I pray you.'

NIKITA. 'Believe me, I cannot.'

SVETLOVIDOV. 'I do beseech you.'

NIKITA. 'I know no touch of it, my lord.'

SVETLOVIDOV. "Tis as easy as lying; govern these ventages with your finger and thumb, give it breath with your mouth, and it will discourse most eloquent music.'

NIKITA. 'I have not the skill.'

SVETLOVIDOV. 'Why, look you now, how unworthy a thing you make of me. You would play upon me; you would seem to know my stops; you would pluck out the heart of my mystery. Do you think I am easier to be played on than a pipe? Call me what instrument you will, though you can fret me, you cannot play upon me.' [*Roars with laughter and claps.*] Bravo! Encore! Bravo! Not much old age about that, was there, damn it? There's no such thing as old age, that's a lot of nonsense. I feel strength pulsing in every vein—why, this is youth, zest, the spice of life! If you're good enough, Nikita, being old doesn't count. Think I'm crazy, eh? Gone off my head, have I? Wait, let me pull myself together. Good Lord above us! Now, listen to this—how tender and subtle. Ah, the music of it! Shush, be quiet!

'O silent night in old Ukraine!
The stars are bright, clear is the sky.
All drowsy is the heavy air
And silver poplars faintly sigh.'

[*There is the sound of doors being opened.*] What's that?

NIKITA. It must be George and Peter. You're good, Mr. Svetlovidov, a great actor!

SVETLOVIDOV [*shouts in the direction of the noise*]. This way, lads! [*To* NIKITA.] Let's go and change. There's no such thing as old age, that's all stuff and nonsense. [*Roars with happy laughter.*] So why the tears? Why so down in the mouth, you dear, silly fellow? Now this won't do, it really won't. Really, old chap, you mustn't look like that—what good does it do? There, there. [*Embraces him with tears in his eyes.*] You mustn't cry. Where art and genius are, there's no room for old age, loneliness and illness—why, death itself loses half its sting. [*Weeps.*] Ah well, Nikita, we've made our last bow. I'm no great actor, just a squeezed lemon, a miserable nonentity, a rusty old nail. And you're just an old stage hack, a prompter. Come on. [*They start to move off.*] I'm no real good—in a serious play I could just about manage a member of Fortinbras's suite, and for that I'm too old. Ah, well. Remember that bit of *Othello*, Nikita?

> 'Farewell the tranquil mind; farewell content!
> Farewell the plumed troop and the big wars
> That make ambition virtue! O, farewell!
> Farewell the neighing steed, and the shrill trump,
> The spirit-stirring drum, the ear-piercing fife,
> The royal banner, and all quality,
> Pride, pomp, and circumstance of glorious war!'

NIKITA. Terrific! Great stuff!

SVETLOVIDOV. Or take this:

> 'Away from Moscow! Never to return!
> I'll flee the place with not a backward glance,
> And scour the globe for some forgotten corner
> To nurse my wounded heart in. Get my coach!'

[*Goes out with* NIKITA.]

THE CURTAIN SLOWLY FALLS

THE BEAR

[*Медведь*]

A FARCE IN ONE ACT

(1888)

(Dedicated to N. N. Solovtsov)

CHARACTERS

MRS. HELEN POPOV, a young widow with dimpled cheeks, a landowner

GREGORY SMIRNOV, a landowner in early middle age

LUKE, Mrs. Popov's old manservant

The action takes place in the drawing-room of Mrs. Popov's country house

SCENE I

[MRS. POPOV, *in deep mourning, with her eyes fixed on a snapshot, and* LUKE.]

LUKE. This won't do, madam, you're just making your life a misery. Cook's out with the maid picking fruit, every living creature's happy and even our cat knows how to enjoy herself—she's parading round the yard trying to pick up a bird or two. But here you are cooped up inside all day like you was in a convent cell—you never have a good time. Yes, it's true. Nigh on twelve months it is since you last set foot outdoors.

MRS. POPOV. And I'm never going out again, why should I? My life's finished. He lies in his grave, I've buried myself inside these four walls—we're both dead.

LUKE. There you go again! I don't like to hear such talk, I don't. Your husband died and that was that—God's will be done and may he rest in peace. You've shed a few tears and that'll do, it's time to call it a day—you can't spend your whole life a-moaning and a-groaning. The same thing happened to me once, when my old woman died, but what did I do? I grieved a bit, shed a tear or two for a month or so and that's all she's getting. Catch me wearing sackcloth and ashes for the rest of my days, it'd be more than the old girl was worth! [*Sighs.*] You've neglected all the neighbours—won't go and see them or have them in the house. We never get out and about, lurking here like dirty great spiders, saving your presence. The mice have been at my livery too. And it's not for any lack of nice people either—the county's full of 'em, see. There's the regiment stationed at Ryblovo and them officers are a fair treat, a proper sight for sore eyes they are. They have a dance in camp of a Friday and the brass band plays most days. This ain't right, missus. You're young, and pretty as a picture with that peaches-and-cream look, so make the most of it. Them looks won't last for ever, you know. If you wait another ten years to come out of your shell and lead them officers a dance, you'll find it's too late.

MRS. POPOV [*decisively*]. Never talk to me like that again, please. When Nicholas died my life lost all meaning, as you know. You may think I'm alive, but I'm not really. I swore to wear this mourning and shun society till my dying day, do you hear? Let his departed spirit see how I love him! Yes, I realize you know what went on—that he was often mean to me, cruel and, er, unfaithful even, but I'll be true to the grave and show him how much I can love. And he'll find me in the next world just as I was before he died.

LUKE. Don't talk like that—walk round the garden instead. Or else have Toby or Giant harnessed and go and see the neighbours.

MRS. POPOV. Oh dear! [*Weeps.*]

LUKE. Missus! Madam! What's the matter? For heaven's sake!

MRS. POPOV. He was so fond of Toby—always drove him when he went over to the Korchagins' place and the Vlasovs'. He drove so well too! And he looked so graceful when he pulled hard on the reins, remember? Oh Toby, Toby! See he gets an extra bag of oats today.

LUKE. Very good, madam.

[*A loud ring.*]

MRS. POPOV [*shudders*]. Who is it? Tell them I'm not at home.

LUKE. Very well, madam. [*Goes out.*]

SCENE II

[MRS. POPOV, *alone.*]

MRS. POPOV [*looking at the snapshot*]. Now you shall see how I can love and forgive, Nicholas. My love will only fade when I fade away myself, when this poor heart stops beating. [*Laughs, through tears.*] Well, aren't you ashamed of yourself? I'm your good, faithful little wifie, I've locked myself up and I'll be faithful to the grave, while you—aren't you ashamed, you naughty boy? You deceived me and you used to make scenes and leave me alone for weeks on end.

SCENE III

[MRS. POPOV *and* LUKE.]

LUKE [*comes in, agitatedly*]. Someone's asking for you, madam. Wants to see you——

MRS. POPOV. Then I hope you told them I haven't received visitors since the day my husband died.

LUKE. I did, but he wouldn't listen—his business is very urgent, he says.

MRS. POPOV. *I am not at home!*

LUKE. So I told him, but he just swears and barges straight in, drat him. He's waiting in the dining-room.

MRS. POPOV [*irritatedly*]. All right, ask him in here then. Aren't people rude?

[LUKE *goes out.*]

MRS. POPOV. Oh, aren't they all a bore? What do they want with me, why must they disturb my peace? [*Sighs.*] Yes, I see I really shall have to get me to a nunnery. [*Reflects.*] I'll take the veil, that's it.

SCENE IV

[MRS. POPOV, LUKE *and* SMIRNOV.]

SMIRNOV [*coming in, to* LUKE]. You're a fool, my talkative friend. An ass. [*Seeing* MRS. POPOV, *with dignity.*] May I introduce myself, madam? Gregory Smirnov, landed gentleman and lieutenant of artillery retired. I'm obliged to trouble you on most urgent business.

MRS. POPOV [*not holding out her hand*]. What do you require?

SMIRNOV. I had the honour to know your late husband. He died owing me twelve hundred roubles—I have his two IOUs. Now I've some interest due to the land-bank tomorrow, madam, so may I trouble you to let me have the money today?

MRS. POPOV. Twelve hundred roubles—. How did my husband come to owe you that?

SMIRNOV. He used to buy his oats from me.

MRS. POPOV [*sighing, to* LUKE]. Oh yes—Luke, don't forget to see Toby has his extra bag of oats. [LUKE *goes out. To* SMIRNOV.] Of course I'll pay if Nicholas owed you something, but I've nothing on me today, sorry. My manager will be back from town the day after tomorrow and I'll get him to pay you whatever it is then, but for the time being I can't oblige. Besides, it's precisely seven months today since my husband died and I am in no fit state to discuss money.

SMIRNOV. Well, I'll be in a fit state to go bust with a capital B if I can't pay that interest tomorrow. They'll have the bailiffs in on me.

MRS. POPOV. You'll get your money the day after tomorrow.

SMIRNOV. I don't want it the day after tomorrow, I want it now.

MRS. POPOV. I can't pay you now, sorry.

SMIRNOV. And I can't wait till the day after tomorrow.

MRS. POPOV. Can I help it if I've no money today?

SMIRNOV. So you can't pay then?

MRS. POPOV. Exactly.

SMIRNOV. I see. And that's your last word, is it?

MRS. POPOV. It is.

SMIRNOV. Your last word? You really mean it?

MRS. POPOV. I do.

SMIRNOV [*sarcastic*]. Then I'm greatly obliged to you, I'll put it in my diary! [*Shrugs.*] And people expect me to be cool and collected! I met the local excise man on my way here just now. 'My dear Smirnov,' says he, 'why are you always losing your temper?' But how can I help it, I ask you? I'm in desperate need of money! Yesterday morning I left home at crack of dawn. I call on everyone who owes me money, but not a soul forks out. I'm dog-tired. I spend the night in some God-awful place—by the vodka barrel in a Jewish pot-house. Then I fetch up here, fifty miles from home, hoping to see the colour of my money, only to be fobbed off with this 'no fit state' stuff! How *can* I keep my temper?

MRS. POPOV. I thought I'd made myself clear. You can have your money when my manager gets back from town.

SMIRNOV. It's not your manager I'm after, it's you. What the blazes, pardon my language, do I want with your manager?

MRS. POPOV. I'm sorry, my dear man, but I'm not accustomed to these peculiar expressions and to this tone. I have closed my ears. [*Hurries out.*]

SCENE V

[SMIRNOV, *alone.*]

SMIRNOV. Well, what price that! 'In no fit state'! Her husband died seven months ago, if you please! Now have I got my interest to pay or not? I want a straight answer—yes or no? All right, your husband's dead, you're in no fit state and so on and so forth, and your blasted manager's hopped it. But what am I supposed to do? Fly away from my creditors by balloon, I take it! Or go and bash the old brain-box against a brick wall? I call on Gruzdev—not at home. Yaroshevich is in hiding. I have a real old slanging-match with Kuritsyn and almost chuck him out of the window. Mazutov has the belly-ache, and this creature's 'in no fit state'. Not one of the swine will pay. This is what comes of being too nice to them and behaving like some snivelling no-hoper or old woman. It doesn't pay to wear kid gloves with this lot! All right, just you wait—I'll give you something to remember me by! You don't make a monkey out of me, blast you! I'm staying here—going to stick around till she coughs up. Pah! I feel well and truly riled today. I'm shaking like a leaf, I'm so furious—choking I am. Phew, my God, I really think I'm going to pass out! [*Shouts.*] Hey, you there!

SCENE VI

[SMIRNOV *and* LUKE.]

LUKE [*comes in*]. What is it?

SMIRNOV. Bring me some kvass or water, will you?

[LUKE *goes out.*]

SMIRNOV. What a mentality, though! You need money so bad you could shoot yourself, but she won't pay, being 'in no fit state to discuss money', if you please! There's female logic for you and no mistake! That's why I don't like talking to women. Never have. Talk to a woman—why, I'd rather sit on top of a powder magazine! Pah! It makes my flesh creep, I'm so fed up with her, her and that great trailing dress! Poetic creatures they call 'em! Why, the very sight of one gives me cramp in both legs, I get so aggravated.

SCENE VII

[SMIRNOV *and* LUKE.]

LUKE [*comes in and serves some water*]. Madam's unwell and won't see anyone.

SMIRNOV. You clear out!

[LUKE *goes out.*]

SMIRNOV. 'Unwell and won't see anyone.' All right then, don't! I'm staying put, chum, and I don't budge one inch till you unbelt. Be ill for a week and I'll stay a week, make it a year and a year I'll stay. I'll have my rights, lady! As for your black dress and dimples, you don't catch me that way—we know all about those dimples! [*Shouts through the window.*] Unhitch, Simon, we're here for some time—I'm staying put. Tell the stable people to give my horses oats. And you've got that animal tangled in the reins again, you great oaf! [*Imitates him.*] 'I don't care.' I'll give you don't care! [*Moves away from the window.*] How ghastly—it's unbearably hot, no one will pay up, I had a bad night, and now here's this female with her long black dress and her states. I've got a headache. How about a glass of vodka? That might be an idea. [*Shouts.*] Hey, you there!

LUKE [*comes in*]. What is it?

SMIRNOV. Bring me a glass of vodka.

[LUKE *goes out.*]

SMIRNOV. Phew! [*Sits down and looks himself over.*] A fine specimen I am, I must say—dust all over me, my boots dirty, unwashed, hair unbrushed, straw on my waistcoat. I bet the little woman took me for a burglar. [*Yawns.*] It's not exactly polite to turn up in a drawing-room in this rig! Well, anyway, I'm not a guest here, I'm collecting money. And there's no such thing as correct wear for the well-dressed creditor.

LUKE [*comes in and gives him the vodka*]. This is a liberty, sir.

SMIRNOV [*angrily*]. What!

LUKE. I, er, it's all right, I just——

SMIRNOV. Who do you think you're talking to? You hold your tongue!

LUKE [*aside*]. Now we'll never get rid of him, botheration take it! It's an ill wind brought him along.

[LUKE *goes out.*]

SMIRNOV. Oh, I'm so furious! I could pulverize the whole world, I'm in such a rage. I feel quite ill. [*Shouts.*] Hey, you there!

SCENE VIII

[MRS. POPOV and SMIRNOV.]

MRS. POPOV [*comes in, with downcast eyes*]. Sir, in my solitude I have grown unaccustomed to the sound of human speech, and I can't stand shouting. I must urgently request you not to disturb my peace.

SMIRNOV. Pay up and I'll go.

MRS. POPOV. As I've already stated quite plainly, I've no ready cash. Wait till the day after tomorrow.

SMIRNOV. I've also had the honour of stating quite plainly that I need the money today, not the day after tomorrow. If you won't pay up now, I'll have to put my head in a gas-oven tomorrow.

MRS. POPOV. Can I help it if I've no cash in hand? This is all rather odd.

SMIRNOV. So you won't pay up now, eh?

MRS. POPOV. I can't.

SMIRNOV. In that case I'm not budging, I'll stick around here till I do get my money. [*Sits down.*] You'll pay the day after tomorrow, you say? Very well, then I'll sit here like this till the day after tomorrow. I'll just stay put exactly as I am. [*Jumps up.*] I ask you—have I got that interest to pay tomorrow or haven't I? Think I'm trying to be funny, do you?

MRS. POPOV. Kindly don't raise your voice at me, sir—we're not in the stables.

SMIRNOV. I'm not discussing stables, I'm asking whether my interest falls due tomorrow. Yes or no?

MRS. POPOV. You don't know how to treat a lady.

SMIRNOV. Oh yes I do.

MRS. POPOV. Oh no you don't. You're a rude, ill-bred person. Nice men don't talk to ladies like that.

SMIRNOV. Now, this *is* a surprise! How do you want me to talk then? In French, I suppose? [*In an angry, simpering voice.*] *Madame, je voo pree.* You won't pay me—how perfectly delightful. Oh, *pardong*, I'm sure—sorry you were troubled! Now isn't the weather divine today? And that black dress looks too, too charming! [*Bows and scrapes.*]

MRS. POPOV. That's silly. And not very clever.

SMIRNOV [*mimics her*]. 'Silly, not very clever.' I don't know how to treat a lady, don't I? Madam, I've seen more women in my time than you have house-sparrows. I've fought three duels over women. There have been twenty-one women in my life. Twelve times it was me broke it off, the other nine got in first. Oh yes! Time was I made an ass of myself, slobbered, mooned around, bowed and scraped and practically crawled on my belly. I loved, I suffered, I sighed at the moon, I languished, I melted, I grew cold. I loved passionately, madly, in every conceivable fashion, damn me, burbling nineteen to the dozen about women's emancipation and wasting half my substance on the tender passion. But now—no thank you very much! I can't be fooled any more, I've had enough. Black eyes, passionate looks, crimson lips, dimpled cheeks, moonlight, 'Whispers, passion's bated breathing'—I don't give a tinker's cuss for the lot now, lady. Present company excepted, all women, large or small, are simpering, mincing, gossipy creatures. They're great haters. They're eyebrow-deep in lies. They're futile, they're trivial, they're cruel, they're outrageously illogical. And as for having anything upstairs [*taps his forehead*]—I'm sorry to be so blunt, but the very birds in the trees can run rings round your average bluestocking. Take any one of these poetical creations. Oh, she's all froth and fluff, she is, she's half divine, she sends you into a million raptures. But you take a peep inside her mind, and what do you see? A common or garden crocodile! [*Clutches the back of a chair, which cracks and breaks.*] And yet this crocodile somehow thinks its great life-work, privilege and monopoly is the tender passion—that's what really gets me! But damn and blast it, and crucify me upside down on that wall if I'm wrong—does a woman know how to love any living creature apart from lap-dogs? Her love gets no further than snivelling and slobbering. The man suffers and makes sacrifices, while she just twitches the train of her dress and tries to get him squirming under her thumb, that's what her love adds up to! You must know what women are like, seeing you've the rotten luck to be one. Tell me frankly, did you ever see a sincere, faithful, true woman? You know you didn't. Only the old and ugly ones are true and faithful. You'll never find a constant woman, not in a month of Sundays you won't, not once in a blue moon!

MRS. POPOV. Well, I like that! Then who is true and faithful in love to your way of thinking? Not men by any chance?

SMIRNOV. Yes, madam. Men.

MRS. POPOV. *Men!* [*Gives a bitter laugh.*] Men true and faithful in love! That's rich, I must say. [*Vehemently.*] What right have you to talk like that? Men true and faithful! If it comes to that, the best man I've ever known was my late husband, I may say. I loved him passionately, with all my heart as only an intelligent young woman can. I gave him my youth, my happiness, my life, my possessions. I lived only for him. I worshipped him as an idol. And—what do you think? This best of men was shamelessly deceiving me all along the line! After his death I found a drawer in his desk full of love letters, and when he was alive—oh, what a frightful memory!—he used to leave me on my own for weeks on end, he carried on with other girls before my very eyes, he was unfaithful to me, he spent my money like water, and he joked about my feelings for him. But I loved him all the same, and I've been faithful to him. What's more, I'm still faithful and true now that he's dead. I've buried myself alive inside these four walls and I shall go round in these widow's weeds till my dying day.

SMIRNOV [*with a contemptuous laugh*]. Widow's weeds! Who do you take me for? As if I didn't know why you wear this fancy dress and bury yourself indoors! Why, it sticks out a mile! Mysterious and romantic, isn't it? Some army cadet or hack poet may pass by your garden, look up at your windows and think: 'There dwells Tamara, the mysterious princess, the one who buried herself alive from love of her husband.' Who do you think you're fooling?

MRS. POPOV [*flaring up*]. *What!* You dare to take that line with me!

SMIRNOV. Buries herself alive—but doesn't forget to powder her nose!

MRS. POPOV. You dare adopt that tone!

SMIRNOV. Don't you raise your voice to me, madam, I'm not one of your servants. Let me call a spade a spade. Not being a woman, I'm used to saying what I think. So stop shouting, pray.

MRS. POPOV. It's you who are shouting, not me. Leave me alone, would you mind?

SMIRNOV. Pay up, and I'll go.

MRS. POPOV. You'll get nothing out of me.

SMIRNOV. Oh yes I shall.

MRS. POPOV. Just to be awkward, you won't get one single copeck. And you can leave me alone.

SMIRNOV. Not having the pleasure of being your husband or fiancé, I'll trouble you not to make a scene. [*Sits down.*] I don't like it.

MRS. POPOV [*choking with rage*]. Do I see you sitting down?

SMIRNOV. You most certainly do.

MRS. POPOV. Would you mind leaving?

SMIRNOV. Give me my money. [*Aside.*] Oh, I'm in such a rage! Furious I am!

MRS. POPOV. I've no desire to bandy words with cads, sir. Kindly clear off! [*Pause.*] Well, are you going or aren't you?

SMIRNOV. No.

MRS. POPOV. No?

SMIRNOV. No!

MRS. POPOV. Very well then! [*Rings.*]

SCENE IX

[*The above and* LUKE.]

MRS. POPOV. Show this gentleman out, Luke.

LUKE [*goes up to* SMIRNOV]. Be so good as to leave, sir, when you're told, sir. No point in——

SMIRNOV [*jumping up*]. You hold your tongue! Who do you think you're talking to? I'll carve you up in little pieces.

LUKE [*clutching at his heart*]. Heavens and saints above us! [*Falls into an armchair.*] Oh, I feel something terrible—fair took my breath away, it did.

MRS. POPOV. But where's Dasha? Dasha! [*Shouts.*] Dasha! Pelageya! Dasha! [*Rings.*]

LUKE. Oh, they've all gone fruit-picking. There's no one in the house. I feel faint. Fetch water.

MRS. POPOV. Be so good as to clear out!

SMIRNOV. Couldn't you be a bit more polite?

MRS. POPOV [*clenching her fists and stamping*]. You uncouth oaf! You have the manners of a bear! Think you own the place? Monster!

SMIRNOV. What! You say that again!

MRS. POPOV. I called you an ill-mannered oaf, a monster!

SMIRNOV [*advancing on her*]. Look here, what right have you to insult me?

MRS. POPOV. All right, I'm insulting you. So what? Think I'm afraid of you?

SMIRNOV. Just because you look all romantic, you can get away with anything—is that your idea? This is duelling talk!

LUKE. Heavens and saints above us! Water!

SMIRNOV. Pistols at dawn!

MRS. POPOV. Just because you have big fists and the lungs of an ox you needn't think I'm scared, see? Think you own the place, don't you!

SMIRNOV. We'll shoot it out! No one calls me names and gets away with it, weaker sex or no weaker sex.

MRS. POPOV [*trying to shout him down*]. You coarse lout!

SMIRNOV. Why should it only be us men who answer for our insults? It's high time we dropped that silly idea. If women want equality, let them damn well have equality! I challenge you, madam!

MRS. POPOV. Want to shoot it out, eh? Very well.

SMIRNOV. This very instant!

MRS. POPOV. Most certainly! My husband left some pistols, I'll fetch them instantly. [*Moves hurriedly off and comes back.*] I'll enjoy putting a bullet through that thick skull, damn your infernal cheek! [*Goes out.*]

SMIRNOV. I'll pot her like a sitting bird. I'm not one of your sentimental young puppies. She'll get no chivalry from me!

LUKE. Kind sir! [*Kneels.*] Grant me a favour, pity an old man and leave this place. First you frighten us out of our wits, now you want to fight a duel.

SMIRNOV [*not listening*]. A duel! There's true women's emancipation for you! That evens up the sexes with a vengeance! I'll knock her off as a matter of principle. But what a woman! [*Mimics her.*] 'Damn your infernal cheek! I'll put a bullet through that thick skull.' Not bad, eh? Flushed all over, flashing eyes, accepts my challenge! You know, I've never seen such a woman in my life.

LUKE. Go away, sir, and I'll say prayers for you till the day I die.

SMIRNOV. There's a regular woman for you, something I do appreciate! A proper woman—not some namby-pamby, wishy-washy female, but a really red-hot bit of stuff, a regular pistol-packing little spitfire. A pity to kill her, really.

LUKE [*weeps*]. Kind sir—do leave. Please!

SMIRNOV. I definitely like her. Definitely! Never mind her dimples, I like her. I wouldn't mind letting her off what she owes me, actually. And I don't feel angry any more. Wonderful woman!

SCENE X

[*The above and* MRS. POPOV.]

MRS. POPOV [*comes in with the pistols*]. Here are the pistols. But before we start would you mind showing me how to fire them? I've never had a pistol in my hands before.

LUKE. Lord help us! Mercy on us! I'll go and find the gardener and coachman. What have we done to deserve this? [*Goes out.*]

SMIRNOV [*examining the pistols*]. Now, there are several types of pistol. There are Mortimer's special duelling pistols with percussion caps. Now, yours here are Smith and Wessons, triple action with extractor, centre-fired. They're fine weapons, worth a cool ninety roubles the pair. Now, you hold a revolver like this. [*Aside.*] What eyes, what eyes! She's hot stuff all right!

MRS. POPOV. Like this?

SMIRNOV. Yes, that's right. Then you raise the hammer and take aim like this. Hold your head back a bit, stretch your arm out properly. Right. And then with this finger you press this little gadget and that's it. But the great thing is—don't get excited and do take your time about aiming. Try and see your hand doesn't shake.

MRS. POPOV. All right. We can't very well shoot indoors, let's go in the garden.

SMIRNOV. Very well. But I warn you, I'm firing in the air.

MRS. POPOV. Oh, this is the limit! Why?

SMIRNOV. Because, because—. That's my business.

MRS. POPOV. Got cold feet, eh? I see. Now don't shilly-shally, sir.

Kindly follow me. I shan't rest till I've put a bullet through your brains, damn you. Got the wind up, have you?

SMIRNOV. Yes.

MRS. POPOV. That's a lie. Why won't you fight?

SMIRNOV. Because, er, because you, er, I like you.

MRS. POPOV [*with a vicious laugh*]. He likes me! He dares to say he likes me! [*Points to the door.*] I won't detain you.

SMIRNOV [*puts down the revolver without speaking, picks up his peaked cap and moves off; near the door he stops and for about half a minute the two look at each other without speaking; then he speaks, going up to her hesitantly*]. Listen. Are you still angry? I'm absolutely furious myself, but you must see—how can I put it? The fact is that, er, it's this way, actually—. [*Shouts.*] Anyway, can I help it if I like you? [*Clutches the back of a chair, which cracks and breaks.*] Damn fragile stuff, furniture! I like you! Do you understand? I, er, I'm almost in love.

MRS. POPOV. Keep away from me, I loathe you.

SMIRNOV. God, what a woman! Never saw the like of it in all my born days. I'm sunk! Without trace! Trapped like a mouse!

MRS. POPOV. Get back or I shoot.

SMIRNOV. Shoot away. I'd die happily with those marvellous eyes looking at me, that's what you can't see—die by that dear little velvet hand. Oh, I'm crazy! Think it over and make your mind up now, because once I leave this place we shan't see each other again. So make your mind up. I'm a gentleman and a man of honour, I've ten thousand a year, I can put a bullet through a coin in mid air and I keep a good stable. Be my wife.

MRS. POPOV [*indignantly brandishes the revolver*]. A duel! We'll shoot it out!

SMIRNOV. I'm out of my mind! Nothing makes any sense. [*Shouts.*] Hey, you there—water!

MRS. POPOV [*shouts*]. We'll shoot it out!

SMIRNOV. I've lost my head, fallen for her like some damfool boy! [*Clutches her hand. She shrieks with pain.*] I love you! [*Kneels.*] I love you as I never loved any of my twenty-one other women—twelve times it was me broke it off, the other nine got in first. But I never loved anyone as much as you. I've gone all sloppy, soft and sentimental. Kneeling like an imbecile, offering my hand! Disgraceful! Scandalous! I haven't been in love for five years, I swore not to, and here I am crashing head over heels, hook, line and sinker! I offer you my hand. Take it or leave it. [*Gets up and hurries to the door.*]

MRS. POPOV. Just a moment.

SMIRNOV [*stops*]. What is it?

MRS. POPOV. Oh, never mind, just go away. But wait. No, go, go away. I hate you. Or no—don't go away. Oh, if you knew how furious I am! [*Throws the revolver on the table.*] My fingers are numb from holding this beastly thing. [*Tears a handkerchief in her anger.*] Why are you hanging about? Clear out!

SMIRNOV. Good-bye.

MRS. POPOV. Yes, yes, go away! [*Shouts.*] Where are you going? Stop. Oh, go away then. I'm so furious! Don't you come near me, I tell you.

SMIRNOV [*going up to her*]. I'm so fed up with myself! Falling in love like a schoolboy! Kneeling down! It's enough to give you the willies! [*Rudely.*] I love you! Oh, it's just what the doctor ordered, this is! There's my interest due in tomorrow, haymaking's upon us—and *you* have to come along! [*Takes her by the waist.*] I'll never forgive myself.

MRS. POPOV. Go away! You take your hands off me! I, er, hate you! We'll sh-shoot it out!

[*A prolonged kiss.*]

SCENE XI

[*The above,* LUKE *with an axe, the gardener with a rake, the coachman with a pitchfork and some workmen with sundry sticks and staves.*]

LUKE [*seeing the couple kissing*]. Mercy on us! [*Pause.*]

MRS. POPOV. [*lowering her eyes*]. Luke, tell them in the stables—Toby gets no oats today.

CURTAIN

THE PROPOSAL

[*Предложение*]

A FARCE IN ONE ACT

(1888–1889)

CHARACTERS

STEPHEN CHUBUKOV, a landowner

NATASHA, his daughter, aged 25

IVAN LOMOV, a landowning neighbour of Chubukov's, hefty and well-nourished, but a hypochondriac

The action takes place in the drawing-room of Chubukov's country-house

SCENE I

[CHUBUKOV *and* LOMOV; *the latter comes in wearing evening dress and white gloves.*]

CHUBUKOV [*going to meet him*]. Why, it's Ivan Lomov—or do my eyes deceive me, old boy? Delighted. [*Shakes hands.*] I say, old bean, this is a surprise! How *are* you?

LOMOV. All right, thanks. And how might you be?

CHUBUKOV. Not so bad, dear boy. Good of you to ask and so on. Now, you simply must sit down. Never neglect the neighbours, old bean—what? But why so formal, old boy—the tails, the gloves and so on? Not going anywhere, are you, dear man?

LOMOV. Only coming here, my dear Chubukov.

CHUBUKOV. Then why the tails, my dear fellow? Why make such a great thing of it?

LOMOV. Well, look, the point is—. [*Takes his arm.*] I came to ask a favour, my dear Chubukov, if it's not too much bother. I have had the privilege of enlisting your help more than once, and you've always, as it were—but I'm so nervous, sorry. I'll drink some water, my dear Chubukov. [*Drinks water.*]

CHUBUKOV [*aside*]. He's come to borrow money. Well, there's nothing doing! [*To him.*] What's the matter, my dear fellow?

LOMOV. Well, you see, my chear Dubukov—my dear Chubukov, I mean, sorry—that's to say, I'm terribly jumpy, as you see. In fact only you can help me, though I don't deserve it, of course and, er, have no claims on you either.

CHUBUKOV. Now don't muck about with it, old bean. Let's have it. Well?

LOMOV. Certainly, this instant. The fact is, I'm here to ask for the hand of your daughter Natasha.

CHUBUKOV [*delightedly*]. My dear Lomov! Say that again, old horse, I didn't quite catch it.

LOMOV. I have the honour to ask——

CHUBUKOV [*interrupting him*]. My dear old boy! I'm delighted and so on, er, and so forth—what? [*Embraces and kisses him.*] I've long wanted it, it's always been my wish. [*Sheds a tear.*] I've always loved you as a son, dear boy. May you both live happily ever after and so on. As for me, I've always wanted—. But why do I stand around like a blithering idiot? I'm tickled pink, I really am! Oh, I most cordially—. I'll go and call Natasha and so forth.

LOMOV [*very touched*]. My dear Chubukov, what do you think—can I count on a favourable response?

CHUBUKOV. What—her turn down a good-looking young fellow like you! Not likely! I bet she's crazy about you and so on. One moment. [*Goes out.*]

SCENE II

[LOMOV, *alone.*]

LOMOV. I feel cold, I'm shaking like a leaf. Make up your mind, that's the great thing. If you keep chewing things over, dithering on the brink, arguing the toss and waiting for your ideal woman or true love to come along, you'll never get hitched up. Brrr! I'm cold. Natasha's a good housewife. She's not bad-looking and she's an educated girl—what more can you ask? But I'm so jumpy, my ears have started buzzing. [*Drinks water.*] And get married I must. In the first place, I'm thirty-five years old—a critical age, so to speak. Secondly, I should lead a proper, regular life. I've heart trouble and constant palpitations, I'm irritable and nervous as a kitten. See how my lips are trembling now? See my right eyelid twitch? But my nights are the worst thing. No sooner do I get in bed and start dozing off than I have a sort of shooting pain in my left side. It goes right through my shoulder and head. Out I leap like a lunatic, walk about a bit, then lie down again—but the moment I start dropping off I get this pain in my side again. And it happens twenty times over.

SCENE III

[NATASHA *and* LOMOV.]

NATASHA [*comes in*]. Oh, it's you. That's funny, Father said it was a dealer collecting some goods or something. Good morning, Mr. Lomov.

LOMOV. And good morning to you, my dear Miss Chubukov.

NATASHA. Excuse my apron, I'm not dressed for visitors. We've been shelling peas—we're going to dry them. Why haven't you been over for so long? Do sit down. [*They sit.*] Will you have lunch?

LOMOV. Thanks, I've already had some.

NATASHA. Or a smoke? Here are some matches. It's lovely weather, but it rained so hard yesterday—the men were idle all day. How much hay have you cut? I've been rather greedy, you know—I mowed all mine, and now I'm none too happy in case it rots. I should have hung on. But what's this I see? Evening dress, it seems. That *is* a surprise! Going dancing or something? You're looking well, by the way—but why on earth go round in that get-up?

LOMOV [*agitated*]. Well, you see, my dear Miss Chubukov. The fact is, I've decided to ask you to—er, lend me your ears. You're bound to be surprised—angry, even. But I—. [*Aside.*] I feel terribly cold.

NATASHA. What's up then? [*Pause.*] Well?

LOMOV. I'll try to cut it short. Miss Chubukov, you are aware that I have long been privileged to know your family—since I was a boy, in fact. My dear departed aunt and her husband—from whom, as you are cognizant, I inherited the estate—always entertained the deepest respect for your father and dear departed mother. We Lomovs and Chubukovs have always been on the friendliest terms—you might say we've been pretty thick. And what's more, as you are also aware, we own closely adjoining properties. You may recall that my land at Oxpen Field is right next to your birch copse.

NATASHA. Sorry to butt in, but you refer to Oxpen Field as 'yours'? Surely you're not serious!

LOMOV. I am, madam.

NATASHA. Well, I like that! Oxpen Field is ours, it isn't yours.

LOMOV. You're wrong, my dear Miss Chubukov, that's my land.

NATASHA. This is news to me. How can it be yours?

LOMOV. How? What do you mean? I'm talking about Oxpen Field, that wedge of land between your birch copse and Burnt Swamp.

NATASHA. That's right. It's our land.

LOMOV. No, you're mistaken, my dear Miss Chubukov. It's mine.

NATASHA. Oh, come off it, Mr. Lomov. How long has it been yours?

LOMOV. How long? As long as I can remember—it's always been ours.

NATASHA. I say, this really is a bit steep!

LOMOV. But you have only to look at the deeds, my dear Miss Chubukov. Oxpen Field once *was* in dispute, I grant you, but it's mine now—that's common knowledge, and no argument about it. If I may explain, my aunt's grandmother made over that field rent free to your father's grandfather's labourers for their indefinite use in return for firing her bricks. Now, your great-grandfather's people used the place rent free for forty years or so, and came to look on it as their own. Then when the government land settlement was brought out——

NATASHA. No, that's all wrong. My grandfather and great-grandfather both claimed the land up to Burnt Swamp as theirs. So Oxpen Field was ours. Why argue? That's what I can't see. This is really rather aggravating.

LOMOV. I'll show you the deeds, Miss Chubukov.

NATASHA. Oh, you must be joking or having me on. This *is* a nice surprise! You own land for nearly three hundred years, then someone ups and tells you it's not yours! Mr. Lomov, I'm sorry, but I simply can't believe my ears. I don't mind about the field—it's only the odd twelve acres, worth the odd three hundred roubles. But it's so unfair—that's what infuriates me. I can't stand unfairness, I don't care what you say.

LOMOV. Do hear me out, please! With due respect, your great-grandfather's people baked bricks for my aunt's grandmother, as I've already told you. Now, my aunt's grandmother wanted to do them a favour——

NATASHA. Grandfather, grandmother, aunt—it makes no sense to me. The field's ours, and that's that.

LOMOV. It's mine.

NATASHA. It's ours! Argue till the cows come home, put on tail-coats by the dozen for all I care—it'll still be ours, ours, ours! I'm not after your property, but I don't propose losing mine either, and I don't care what you think!

LOMOV. My dear Miss Chubukov, it's not that I need that field—it's the principle of the thing. If you want it, have it. Take it as a gift.

NATASHA. But it's mine to give *you* if I want—it's my property. This is odd, to put it mildly. We always thought you such a good neighbour and friend, Mr. Lomov. We lent you our threshing-machine last year, and couldn't get our own threshing done till November in consequence. We might be gipsies, the way you treat us. Making me a present of my own property! I'm sorry, but that's not exactly neighbourly of you. In fact, if you ask me, it's sheer howling cheek.

LOMOV. So I'm trying to pinch your land now, am I? It's not my habit, madam, to grab land that isn't mine, and I won't have anyone say it is! [*Quickly goes to the carafe and drinks some water.*] Oxpen Field belongs to me.

NATASHA. That's a lie, it's ours.

LOMOV. It's mine.

NATASHA. That's a lie and I'll nail it! I'll send my men to cut that field this very day.

LOMOV. What do you say?

NATASHA. My men will be out on that field today!

LOMOV. Too right, they'll be out! Out on their ear!

NATASHA. You'd never dare.

LOMOV [*clutches his heart*]. Oxpen Field belongs to me, do you hear? It's mine!

NATASHA. Kindly stop shouting. By all means yell yourself blue in the face when you're in your own home, but I'll thank you to keep a civil tongue in your head in this house.

LOMOV. Madam, if I hadn't got these awful, agonizing palpitations and this throbbing in my temples, I'd give you a piece of my mind! [*Shouts.*] Oxpen Field belongs to me.

NATASHA. To us, you mean!

LOMOV. It's mine!

NATASHA. It's ours!

LOMOV. Mine!

SCENE IV

[*The above and* CHUBUKOV.]

CHUBUKOV [*coming in*]. What's going on, what's all the row in aid of?

NATASHA. Father, who owns Oxpen Field? Would you mind telling this gentleman? Is it his or ours?

CHUBUKOV [*to* LOMOV]. That field's ours, old cock!

LOMOV. Now look here, Chubukov, how can it be? You at least might show some sense! My aunt's grandmother made over that field to your grandfather's farm-labourers rent free on a temporary basis. Those villagers had the use of the land for forty years and came to think of it as theirs, but when the settlement came out——

CHUBUKOV. Now hang on, dear man, you forget one thing. That field was in dispute and so forth even in those days—and that's why the villagers paid your grandmother no rent and so on. But now it belongs to us, every dog in the district knows that, what? You can't have seen the plans.

LOMOV. It's mine and I'll prove it.

CHUBUKOV. Oh no you won't, my dear good boy.

LOMOV. Oh yes, I will.

CHUBUKOV. No need to shout, old bean. Shouting won't prove anything, what? I'm not after your property, but I don't propose losing mine, either. Why on earth should I? If it comes to that, old sausage, if you're set on disputing the field and so on, I'd rather give it to the villagers than you. So there.

LOMOV. This makes no sense to me. What right have you to give other people's property away?

CHUBUKOV. Permit me to be the best judge of that. Now, look here, young feller-me-lad—I'm not used to being spoken to like this, what? I'm twice your age, boy, and I'll thank you to talk to me without getting hot under the collar and so forth.

LOMOV. Oh, really, you must take me for a fool. You're pulling my leg. You say my land's yours, then you expect me to keep my temper and talk things over amicably. I call this downright unneighbourly, Chubukov. You're not a neighbour, you're a thoroughgoing shark!

CHUBUKOV. I *beg* your pardon! What did you say?

NATASHA. Father, send the men out to mow that field this very instant!

CHUBUKOV [*to* LOMOV]. What was it you said, sir?

NATASHA. Oxpen Field's ours and I won't let it go, I won't, I won't!

LOMOV. We'll see about that! I'll have the law on you!

CHUBUKOV. You will, will you? Then go right ahead, sir, and so forth, go ahead and sue, sir! Oh, I know your sort! Just what you're angling for and so on, isn't it—a court case, what? Quite the legal eagle, aren't you? Your whole family's always been litigation-mad, every last one of 'em!

LOMOV. I'll thank you not to insult my family. We Lomovs have always been honest, we've none of us been had up for embezzlement like your precious uncle.

CHUBUKOV. The Lomovs have always been mad as hatters!

NATASHA. Yes! All of you! Mad!

CHUBUKOV. Your grandfather drank like a fish, and your younger Aunt What's-her-name—Nastasya—ran off with an architect and so on.

LOMOV. And your mother was a cripple. [*Clutches his heart.*] There's that shooting pain in my side, and a sort of blow on the head. Heavens alive! Water!

CHUBUKOV. Your father gambled and ate like a pig!

NATASHA. Your aunt was a most frightful busybody!

LOMOV. My left leg's gone to sleep. And you're a very slippery customer. Oh my heart! And it's common knowledge that at election time you bri—. I'm seeing stars. Where's my hat?

NATASHA. What a rotten, beastly, filthy thing to say.

CHUBUKOV. You're a thoroughly nasty, cantankerous, hypocritical piece of work, what? Yes, sir!

LOMOV. Ah, there's my hat. My heart—. Which way do I go? Where's the door? Oh, I think I'm dying. I can hardly drag one foot after another. [*Moves to the door.*]

CHUBUKOV [*after him*]. You need never set either of those feet in my house again, sir.

NATASHA. Go ahead and sue, we'll see what happens.

[LOMOV *goes out staggering.*]

SCENE V

[CHUBUKOV *and* NATASHA.]

CHUBUKOV. Oh, blast it! [*Walks up and down in agitation.*]

NATASHA. The rotten cad! So much for trusting the dear neighbours!

CHUBUKOV. Scruffy swine!

NATASHA. He's an out-and-out monster! Pinches your land and then has the cheek to swear at you!

CHUBUKOV. And this monstrosity, this blundering oaf, has the immortal rind to come here with his proposal and so on, what? A proposal! I ask you!

NATASHA. A proposal, did you say?

CHUBUKOV. Not half I did! He came here to propose to you!

NATASHA. Propose? To me? Then why didn't you say so before?

CHUBUKOV. That's why he dolled himself up in tails. Damn popinjay! Twerp!

NATASHA. Me? Propose to me? Oh! [*Falls in an armchair and groans.*] Bring him back! Bring him back! Bring him back, I tell you!

CHUBUKOV. Bring who back?

NATASHA. Hurry up, be quick, I feel faint. Bring him back. [*Has hysterics.*]

CHUBUKOV. What's this? What do you want? [*Clutches his head.*] Oh, misery! I might as well go and boil my head! I'm fed up with them!

NATASHA. I'm dying. Bring him back!

CHUBUKOV. Phew! All right then. No need to howl. [*Runs out.*]

NATASHA [*alone, groans*]. What have we done! Bring him, bring him back!

CHUBUKOV [*runs in*]. He'll be here in a moment and so on, damn him! Phew! You talk to him—I don't feel like it, what?

NATASHA [*groans*]. Bring him back!

CHUBUKOV [*shouts*]. He's coming, I tell you.

'My fate, ye gods, is just too bad—
To be a grown-up daughter's dad!'

I'll cut my throat, I'll make a point of it. We've sworn at the man, insulted him and kicked him out of the house. And it was all your doing.

NATASHA. It was *not*, it was yours!

CHUBUKOV. So now it's my fault, what?

[LOMOV *appears in the doorway*.]

CHUBUKOV. All right, now you talk to him. [*Goes out*.]

SCENE VI

[NATASHA *and* LOMOV.]

LOMOV [*comes in, exhausted*]. My heart's fairly thumping away, my leg's gone to sleep and there's this pain in my side——

NATASHA. I'm sorry we got a bit excited, Mr. Lomov. I've just remembered—Oxpen Field really does belong to you.

LOMOV. My heart's fairly thumping away. That field's mine. I've a nervous tic in both eyes.

NATASHA. The field *is* yours, certainly. Do sit down. [*They sit*.] We were mistaken.

LOMOV. This is a question of principle. It's not the land I mind about, it's the principle of the thing.

NATASHA. Just so, the principle. Now let's change the subject.

LOMOV. Especially as I can prove it. My aunt's grandmother gave your father's grandfather's villagers——

NATASHA. All right, that'll do. [*Aside*.] I don't know how to start. [*To him*.] Thinking of going shooting soon?

LOMOV. Yes, I'm thinking of starting on the woodcock after the harvest, my dear Miss Chubukov. I say, have you heard? What awful bad luck! You know my dog Tracker? He's gone lame.

NATASHA. Oh, I am sorry. How did it happen?

LOMOV. I don't know. Either it must be a sprain, or the other dogs bit him. [*Sighs*.] My best dog, to say nothing of what he set me back!

Do you know, I gave Mironov a hundred and twenty-five roubles for him?

NATASHA. Then you were had, Mr. Lomov.

LOMOV. He came very cheap if you ask me—he's a splendid dog.

NATASHA. Father only gave eighty-five roubles for Rover. And Rover's a jolly sight better dog than Tracker, you'll agree.

LOMOV. Rover better than Tracker! Oh, come off it! [*Laughs.*] Rover a better dog than Tracker!

NATASHA. Of course he is. Rover's young, it's true, and not yet in his prime. But you could search the best kennels in the county without finding a nippier animal, or one with better points.

LOMOV. I am sorry, Miss Chubukov, but you forget he has a short lower jaw, and a dog like that can't grip.

NATASHA. Oh, can't he! That's news to me!

LOMOV. He has a weak chin, you can take that from me.

NATASHA. Why, have you measured it?

LOMOV. Yes, I have. Naturally he'll do for coursing, but when it comes to retrieving, that's another story.

NATASHA. In the first place, Rover has a good honest coat on him, and a pedigree as long as your arm. As for that mud-coloured, piebald animal of yours, his antecedents are anyone's guess, quite apart from him being ugly as a broken-down old cart-horse.

LOMOV. Old he may be, but I wouldn't swap him for half a dozen Rovers—not on your life! Tracker's a real dog, and Rover—why, it's absurd to argue. The kennels are lousy with Rovers, he'd be dear at twenty-five roubles.

NATASHA. You *are* in an awkward mood today, Mr. Lomov. First you decide our field is yours, now you say Tracker's better than Rover. I dislike people who won't speak their mind. Now, you know perfectly well that Rover's umpteen times better than that—yes, that stupid Tracker. So why say the opposite?

LOMOV. I see you don't credit me with eyes or brains, Miss Chubukov. Well, get it in your head that Rover has a weak chin.

NATASHA. That's not true.

LOMOV. Oh yes it is!

NATASHA [*shouts*]. Oh no it isn't!

LOMOV. Don't you raise your voice at me, madam.

NATASHA. Then don't you talk such utter balderdash! Oh, this is infuriating! It's time that measly Tracker was put out of his misery—and you compare him with Rover!

LOMOV. I can't go on arguing, sorry—it's my heart.

NATASHA. Men who argue most about sport, I've noticed, are always the worst sportsmen.

LOMOV. Will you kindly hold your trap, madam—my heart's breaking in two. [*Shouts.*] You shut up!

NATASHA. I'll do nothing of the sort till you admit Rover's a hundred times better than Tracker.

LOMOV. A hundred times worse, more like! I hope Rover drops dead! Oh, my head, my eye, my shoulder——

NATASHA. That half-wit Tracker doesn't need to drop dead—he's pretty well a walking corpse already.

LOMOV [*weeps*]. Shut up! I'm having a heart attack!

NATASHA. I will *not* shut up!

SCENE VII

[*The above and* CHUBUKOV.]

CHUBUKOV [*comes in*]. What is it this time?

NATASHA. Father, I want an honest answer: which is the better dog, Rover or Tracker?

LOMOV. Will you kindly tell us just one thing, Chubukov: has Rover got a weak chin or hasn't he? Yes or no?

CHUBUKOV. What if he has? As if that mattered! Seeing he's only the best dog in the county and so on.

LOMOV. Tracker's better, and you know it! Be honest!

CHUBUKOV. Keep your shirt on, dear man. Now look here. Tracker has got some good qualities, what? He's a pedigree dog, has firm paws, steep haunches and so forth. But that dog has two serious faults if you want to know, old bean: he's old and he's pug-nosed.

LOMOV. I'm sorry—it's my heart! Let's just look at the facts. You may recall that Tracker was neck and neck with the Count's Swinger on Maruskino Green when Rover was a good half-mile behind.

CHUBUKOV. He dropped back because the Count's huntsman fetched him a crack with his whip.

LOMOV. Serve him right. Hounds are all chasing the fox and Rover has to start worrying a sheep!

CHUBUKOV. That's not true, sir. I've got a bad temper, old boy, and the fact is—let's please stop arguing, what? He hit him because everyone hates the sight of another man's dog. Oh yes they do. Loathe 'em, they do. And you're no one to talk either, sir! The moment you spot a better dog than the wretched Tracker, you always try to start something and, er, so forth—what? I don't forget, you see.

LOMOV. Nor do I, sir.

CHUBUKOV [*mimics him*]. 'Nor do I, sir.' What is it you don't forget then?

LOMOV. My heart! My leg's gone to sleep. I can't go on.

NATASHA [*mimics him*]. 'My heart!' Call yourself a sportsman! You should be lying on the kitchen stove squashing black-beetles, not fox-hunting. His heart!

CHUBUKOV. Some sportsman, I must say! With that heart you should stay at home, not bob around in the saddle, what? I wouldn't mind if you hunted properly, but you only turn out to pick quarrels and annoy the hounds and so on. I have a bad temper, so let's change the subject. You're no sportsman, sir—what?

LOMOV. What about you then? You only turn out so you can get in the Count's good books and intrigue against people. Oh, my heart! You're a slippery customer, sir!

CHUBUKOV. What's that, sir? Oh, I am, am I? [*Shouts.*] Hold your tongue!

LOMOV. You artful old dodger!

CHUBUKOV. Why, you young puppy!

LOMOV. Nasty old fogy! Canting hypocrite!

CHUBUKOV. Shut up, or I'll pot you like a ruddy partridge. And I'll use a dirty gun too, you idle gasbag!

LOMOV. And it's common knowledge that—oh, my heart—your wife used to beat you. Oh, my leg! My head! I can see stars! I can't stand up!

CHUBUKOV. And your housekeeper has you eating out of her hand!

LOMOV. Oh, oh! My heart's bursting. My shoulder seems to have come off—where is the thing? I'm dying. [*Falls into an armchair.*] Fetch a doctor. [*Faints.*]

CHUBUKOV. Why, you young booby! Hot air merchant! I think I'm going to faint. [*Drinks water.*] I feel unwell.

NATASHA. Calls himself a sportsman and can't even sit on a horse! [*To her father.*] Father, what's the matter with him? Father, have a look. [*Screeches.*] Mr. Lomov! He's dead!

CHUBUKOV. I feel faint. I can't breathe! Give me air!

NATASHA. He's dead. [*Tugs* LOMOV'*s sleeve.*] Mr. Lomov, Mr. Lomov! What have we done? He's dead. [*Falls into an armchair.*] Fetch a doctor, a doctor! [*Has hysterics.*]

CHUBUKOV. Oh! What's happened? What's the matter?

NATASHA [*groans*]. He's dead! Dead!

CHUBUKOV. Who's dead? [*Glancing at* LOMOV.] My God, you're right! Water! A doctor! [*Holds a glass to* LOMOV'*s mouth.*] Drink! No, he's not drinking. He must be dead, and so forth. Oh, misery, misery! Why don't I put a bullet in my brain? Why did I never get round to cutting my throat? What am I waiting for? Give me a knife! A pistol! [LOMOV *makes a movement.*] I think he's coming round. Drink some water! That's right.

LOMOV. I can see stars! There's a sort of mist. Where am I?

CHUBUKOV. Hurry up and get married and—oh, to hell with you! She says yes. [*Joins their hands.*] She says yes, and so forth. You have my blessing, and so on. Just leave me in peace, that's all.

LOMOV. Eh? What? [*Raising himself.*] Who?

CHUBUKOV. She says yes. Well, what about it? Kiss each other and—oh, go to hell!

NATASHA [*groans*]. He's alive. Yes, yes, yes! I agree.

CHUBUKOV. Come on, kiss.

LOMOV. Eh? Who? [*Kisses* NATASHA.] Very nice too. I say, what's all this about? Oh, I see—. My heart! I'm seeing stars! Miss Chubukov, I'm so happy. [*Kisses her hand.*] My leg's gone to sleep.

NATASHA. I, er, I'm happy too.

CHUBUKOV. Oh, what a weight off my mind! Phew!

NATASHA. Still, you must admit now that Tracker's not a patch on Rover.

LOMOV. Oh yes he is!

NATASHA. Oh no he isn't!

CHUBUKOV. You can see those two are going to live happily ever after! Champagne!

LOMOV. He's better.

NATASHA. He's worse, worse, worse.

CHUBUKOV [*trying to shout them down*]. Champagne, champagne, champagne!

CURTAIN

TATYANA REPIN

[*Татьяна Репина*]

A DRAMA IN ONE ACT

(1889)

(Dedicated to A. S. Suvorin)

CHARACTERS

VERA OLENIN

MRS. KOKOSHKIN

MATVEYEV

SONNENSTEIN

PETER SABININ

KOTELNIKOV

KOKOSHKIN

PATRONNIKOV

VOLGIN, a young officer

A student

A young lady

FATHER IVAN, the cathedral dean, aged 70

FATHERS NICHOLAS and ALEXIS, young priests

Deacon

Acolyte

KUZMA, the cathedral caretaker

Woman in black

The assistant public prosecutor

Actors and actresses

Between six and seven o'clock in the evening. The cathedral. All lamps and candles are burning. The holy gates in front of the chancel are open. Two choirs are taking part: the bishop's choir and the cathedral choir. The church is crowded, close and stuffy. A marriage service is being performed. VERA *and* SABININ *are the bride and groom. The groom is attended by* KOTELNIKOV *and* VOLGIN, *the bride by her student brother and the* ASSISTANT PUBLIC PROSECUTOR. *The entire local intelligentsia is present. Smart dresses. The officiating clergy are:* FATHER IVAN *in a faded high hat;* FATHER NICHOLAS, *who wears a low cap and has a lot of hair; and* FATHER ALEXIS, *who is very young and wears dark glasses. In the rear and somewhat to the right of* FATHER IVAN *stands the tall thin* DEACON *with a book in his hands. The congregation includes the local theatrical company headed by* MATVEYEV.

FATHER IVAN [*reading*]. Remember, O God, the parents who have reared them; for the prayers of parents confirm the foundation of houses. Remember, O Lord our God, thy servants, the paranymphs, who are present at this rejoicing. Remember, O Lord our God, thy servant PETER and thine handmaid VERA, and bless them. Give them fruit of the womb, fair children and unanimity of soul and body. Exalt them as the cedars of Libanus, and as a well-cultured vine. Bestow upon them seed of corn, that, having every sufficiency, they may abound in every work that is good and acceptable unto thee; and let them behold their son's sons as newly planted olive-trees round about their table; and, being accepted before thee, may they shine as the luminaries in heaven unto thee, our Lord. And, together with thee, be glory, might, honour, and worship, to thine unbeginning Father, and to thy life-creating Spirit, now and ever, and to ages of ages.

BISHOP'S CHOIR [*singing*]. Amen.

PATRONNIKOV. It's close in here. What's that medal you've got round your neck, Mr. Sonnenstein?

SONNENSTEIN. It's Belgian. But why such a large congregation? Who let them in, isn't it? Ugh! This is worse than a village bath-house!

PATRONNIKOV. It was the wretched police.

DEACON. Let us pray to the Lord.

CATHEDRAL CHOIR [*singing*]. Lord, have mercy.

FATHER NICHOLAS [*reading*]. O Holy God, who didst form man from the dust, and from his rib didst fashion woman, and yoke her unto him a helpmeet for him, because so it was seemly unto thy majesty for man not to be alone upon the earth; do thou thyself now, O Master, stretch forth thy hand from thy holy dwelling-place, and conjoin this thy servant PETER, and this thine handmaid VERA; for by thee a woman is conjoined unto a man. Yoke them together in unanimity, crown them in one flesh, bestow on them fruit of the womb, and the gain of well-favoured children.

For thine is the might, and thine is the kingdom, and the power, and the glory, of the Father, and of the Son, and of the Holy Ghost, now and ever, and to ages of ages.

CATHEDRAL CHOIR [*singing*]. Amen.

YOUNG LADY [*to* SONNENSTEIN]. They're just going to put the crowns on. Look, look!

FATHER IVAN [*taking a crown from the lectern and turning his face to* SABININ]. The servant of God PETER is crowned for the handmaid of God VERA, in the name of the Father, and of the Son, and of the Holy Ghost. Amen. [*Hands the crown to* KOTELNIKOV.]

VOICES IN THE CONGREGATION. 'The best man's just as tall as the groom. He's not much to look at, though. Who is he?' 'That's Kotelnikov. That officer's not up to much either.' 'I say, will you let the lady through.' 'You can't get through here, madam.'

FATHER IVAN [*addresses* VERA]. The handmaid of God VERA is crowned for the servant of God PETER, in the name of the Father, and of the Son, and of the Holy Ghost. [*Hands the crown to the* STUDENT.]

KOTELNIKOV. These crowns are heavy. My hand feels numb already.

VOLGIN. Never mind, I'll take over soon. Who's that stinking of cheap scent? That's what I'd like to know.

ASSISTANT PROSECUTOR. It's Kotelnikov.

KOTELNIKOV. That's a lie.

VOLGIN. Shush!

FATHER IVAN. O Lord our God, crown them with glory and honour.

O Lord our God, crown them with glory and honour. O Lord our God, crown them with glory and honour.

MRS. KOKOSHKIN [*to her husband*]. Doesn't Vera look nice today? I've been admiring her. And she isn't a bit nervous.

KOKOSHKIN. She's used to it. It is her second wedding, after all.

MRS. KOKOSHKIN. Well, that's true enough. [*Sighs.*] I do hope she'll be happy, she has such a kind heart.

ACOLYTE [*coming into the middle of the church*]. The prokimenon of the Epistle, tone viii. Thou hast set upon their heads crowns of precious stones: they asked life of thee, and thou gavest it them.

BISHOP'S CHOIR [*singing*]. Thou hast set upon their heads——

KOTELNIKOV. I wish I could have a smoke.

ACOLYTE. The words of Paul the Apostle.

DEACON. Let us hearken.

ACOLYTE [*intoning in a deep bass*]. Brethren, give thanks always for all things unto God and the Father in the name of our Lord Jesus Christ; submitting yourselves one to another in the fear of God. Wives, submit yourselves unto your own husbands, as unto the Lord. For the husband is the head of the wife, even as Christ is the head of the church: and he is the saviour of the body. Therefore as the church is subject unto Christ, so let the wives be to their own husbands in every thing——

SABININ [*to* KOTELNIKOV]. You're hurting my head with the crown.

KOTELNIKOV. Nonsense, I'm holding it a good six inches above your head.

SABININ. You're squashing my head, I tell you.

ACOLYTE. Husbands, love your wives, even as Christ also loved the church, and gave himself for it; that he might sanctify and cleanse it with the washing of water by the word, that he might present it to himself a glorious church, not having spot, or wrinkle, or any such thing; but that it should be holy and without blemish.

VOLGIN. That's a fine deep voice. [*To* KOTELNIKOV.] Shall I take over?

KOTELNIKOV. I'm not tired yet.

ACOLYTE. So ought men to love their wives as their own bodies. He that loveth his wife loveth himself. For no man ever yet hated his own flesh; but nourisheth and cherisheth it, even as the Lord the church: for we are members of his body, of his flesh, and of his bones. For this cause shall a man leave his father and mother——

SABININ. Hold the crown a bit higher, you're squashing me.

KOTELNIKOV. Rubbish.

ACOLYTE. —and shall be joined unto his wife, and they two shall be one flesh.

KOKOSHKIN. The Governor's here.

MRS. KOKOSHKIN. Where do you see him?

KOKOSHKIN. Over there, standing in front on the right with Mr. Altukhov. He's here unofficially.

MRS. KOKOSHKIN. I see, I see. He's talking to Masha Ganzen. He's crazy about her.

ACOLYTE. This is a great mystery: but I speak concerning Christ and the church. Nevertheless let every one of you in particular so love his wife even as himself; and the wife see that she reverence her husband.

CATHEDRAL CHOIR [*singing*]. Hallelujah, hallelujah, hallelujah.

VOICES. 'Hear that, Natalya? "Let the wife reverence her husband."' 'Oh, you leave me alone!' [*Laughter.*] 'Shush! This won't do, you people.'

DEACON. Wisdom, standing, let us hear the Holy Gospel.

FATHER IVAN. Peace be to all.

CATHEDRAL CHOIR [*singing*]. And to thy spirit.

VOICES. 'Reading the Epistle and the New Testament—how long it all takes! It's time they gave us a rest.' 'You can't breathe in here, I'm going.' 'You won't get through. Wait a bit, it'll be over soon.'

FATHER IVAN. The lesson from the Gospel of Saint John.

ACOLYTE. Let us hearken.

FATHER IVAN [*after taking off his tall hat*]. At that time there was a marriage in Cana of Galilee; and the mother of Jesus was there: and both Jesus was called, and his disciples, to the marriage. And when they wanted wine, the mother of Jesus saith unto him, They have no wine. Jesus saith unto her, Woman, what have I to do with thee? mine hour is not yet come.

SABININ [*to* KOTELNIKOV]. Will it be over soon?

KOTELNIKOV. I don't know, I'm not well up in these things. I should think it must be.

VOLGIN. The bride and groom still have to make their procession.

FATHER IVAN. His mother saith unto the servants, Whatsoever he saith unto you, do it. And there were set there six waterpots of stone, after the manner of the purifying of the Jews, containing two or three firkins apiece. Jesus saith unto them, Fill the waterpots with water. And they filled them up to the brim. And he saith unto them, Draw out now, and bear unto the governor of the feast——

[*A groan is heard.*]

VOLGIN. *Qu'est-ce que c'est*? Has someone been trodden on?

VOICES. 'Shush! Shush!'

[*A groan.*]

FATHER IVAN. And they bare it. When the ruler of the feast had tasted the water that was made wine, and knew not whence it was: (but the servants which drew the water knew;) the governor of the feast called the bridegroom, and saith unto him——

SABININ [*to* KOTELNIKOV]. Who gave that groan just now?

KOTELNIKOV [*staring at the congregation*]. Something's moving there, a woman in black. She must have been taken ill, they're leading her out.

SABININ [*staring*]. Hold the crown a bit higher.

FATHER IVAN. Every man at the beginning doth set forth good wine; and when men have well drunk, then that which is worse: but thou hast kept the good wine until now. This beginning of miracles did Jesus in Cana of Galilee, and manifested forth his glory; and his disciples believed on him——

VOICE. 'I don't know why they let hysterical women in here.'

BISHOP'S CHOIR [*singing*]. Glory be to thee, O Lord, glory be to thee.

PATRONNIKOV. Stop buzzing like a bee, Sonnenstein. And don't stand with your back to the chancel, it's not done.

SONNENSTEIN. It's the young lady who is buzzing like a bee, not me, tee hee hee!

ACOLYTE. Let us all say with our whole soul, and with our whole mind let us say——

CATHEDRAL CHOIR [*singing*]. Lord have mercy.

DEACON. Lord almighty, God our Father, we pray to thee, do thou listen and have mercy.

VOICES. 'Shush! Be quiet!' 'But I'm being pushed myself.'

CHOIR [*sings*]. Lord, have mercy.

VOICES. 'Be quiet! Shush!' 'Who's that fainted?'

DEACON. Have mercy on us, O Lord, in thy great kindness, we pray thee, do thou listen and have mercy.

CHOIR [*sings three times*]. Lord, have mercy.

DEACON. Let us also pray for our Most Pious Autocratic Great Lord, THE EMPEROR ALEXANDER ALEKSANDROVICH of all Russia, for his power, victory, life, peace, health and salvation. May our Lord God help and succour him in all things and humble all his enemies and foes beneath his feet.

CHOIR [*sings three times*]. Lord, have mercy.

[*A groan. Movement in the crowd.*]

MRS. KOKOSHKIN. What's that? [*To the woman standing next to her.*] My dear, this is intolerable. Why can't they open the doors or something? We'll all die of heat.

VOICES. 'They're trying to take her out, but she won't go.' 'Who is it?' 'Shush!'

DEACON. Let us also pray for his Consort, the Most Pious Lady, THE EMPRESS MARIYA FEODOROVNA.

CHOIR [*sings*]. Lord, have mercy.

DEACON. Let us also pray for His Heir, the Right-believing Lord, THE CESAREVITCH and GRAND DUKE NICHOLAS ALEKSANDROVICH and for the whole ruling house.

CHOIR [*sings*]. Lord, have mercy.

SABININ. Oh, my God——

VERA. What is it?

DEACON. Let us also pray for the Most Holy Governing SYNOD and for our most reverend Lord THEOPHILUS, the Bishop of This and That, and all our brothers in Christ.

CHOIR [*sings*]. Lord, have mercy.

VOICES. 'Another woman poisoned herself in the Hotel Europe yesterday.' 'Yes, they say it was some doctor's wife.' 'Why did she do it, do you know?'

DEACON. Let us also pray for all their Christ-loving army.

CHOIR [*sings*]. Lord, have mercy.

VOLGIN. It sounds as if someone's crying. The congregation's behaving disgracefully.

DEACON. Let us also pray for our brethren, priests and monks and all our brothers in Christ.

CHOIR [*sings*]. Lord, have mercy.

MATVEYEV. The choirs are singing well today.

COMIC ACTOR. I wish we could get singers like that, Mr. Matveyev.

MATVEYEV. You don't expect much, do you, funny-face! [*Laughter.*] Shush!

DEACON. Let us also pray for the grace, life, peace, health, happiness and good fortune of the servants of the Lord PETER and VERA.

CHOIR [*sings*]. Lord, have mercy.

DEACON. Let us also pray for the blessed——

VOICE. 'Yes, some doctor's wife—in the hotel.'

DEACON. And for the Most Holy Orthodox Patriarchs of eternal memory——

VOICES. 'That's the fourth one to do a Tatyana Repin and poison herself. How do you explain these poisonings, old man?' 'Sheer neuroticism, what else?' 'Are they copying each other, do you think?'

DEACON. And the Pious Tsars and Right-believing Tsarinas and those who keep this Holy Temple and all fathers and brothers who rest in the Lord——

VOICES. 'Suicide's catching.' 'There are so many neurotic females about these days, it's something awful!' 'Be quiet. And stop moving around.'

DEACON. —Christian people here and everywhere.

VOICE. 'Less shouting, please.'

[*A groan.*]

CHOIR [*sings*]. Lord, have mercy.

VOICES. 'Tatyana's death has poisoned the air. Our ladies have all caught the disease, their grievances have driven them mad.' 'Even the air in church is poisoned. Can you feel the tension?'

DEACON. Let us also pray for those who work fruitfully and do good, labouring, singing and standing, in this holy and reverend temple, awaiting the bounty of thy mighty grace.

CHOIR [*sings*]. Lord, have mercy.

FATHER IVAN. For thou art a merciful God, and lovest mankind, and to thee we ascribe glory—to the Father, and to the Son, and to the Holy Ghost, now and ever, and to ages of ages.

CHOIR [*sings*]. Amen.

SABININ. I say, Kotelnikov.

KOTELNIKOV. What is it?

SABININ. Nothing. Oh God, Tatyana Repin's here. She's here, I tell you!

KOTELNIKOV. You must be crazy.

SABININ. That woman in black—it's her. I recognized her, saw her.

KOTELNIKOV. There's no resemblance. She's just another brunette, that's all.

DEACON. Let us pray unto the Lord.

KOTELNIKOV. Don't whisper, it's bad manners. People are looking at you.

SABININ. For God's sake—I can hardly stand. It's her.

[*A groan.*]

CHOIR. Lord, have mercy.

VOICES. 'Be quiet. Shush! Who's that pushing at the back, you people? Shush!' 'They've taken her behind a pillar.' 'You can't move anywhere for women, they should stay at home.'

SOMEONE [*shouts*]. Be quiet!

FATHER IVAN [*reads*]. O Lord our God, who, in thy saving providence, didst vouchsafe in Cana of Galilee—. [*Looks at the congregation.*]

What people, I must say! [*Continues reading.*]—to declare marriage honourable by thy presence—. [*Raising his voice.*] I must ask you to be quieter. You're interfering with the service. Please don't walk about the church, and don't talk or make a noise, but stand still and pray. That's right. Have fear of the Lord. [*Reads.*] O Lord our God, who, in Thy saving providence, didst vouchsafe in Cana of Galilee to declare marriage honourable by thy presence; do thou now thyself preserve in peace and unanimity thy servants PETER and VERA, whom thou art well pleased should be conjoined to one another: declare their marriage honourable: preserve their bed undefiled: be pleased that their mutual life may be unblameable, and count them worthy to attain unto a ripe old age, keeping thy commandments in a pure heart.

For thou art our God, the God to have mercy and to save, and to thee we ascribe glory, with thine unbeginning Father, and with thine all-holy, and good, and life-creating Spirit, now and ever, and to ages of ages.

BISHOP'S CHOIR [*sings*]. Amen.

SABININ [*to* KOTELNIKOV]. Have the police been told not to let anyone else in the church?

KOTELNIKOV. What do you mean? They're jammed in like sardines already. Be quiet, stop whispering.

SABININ. She—Tatyana's here.

KOTELNIKOV. You're raving. She's dead and buried.

DEACON. Help us, save us, have mercy on us, and keep us, O God, by thy grace.

CATHEDRAL CHOIR [*singing*]. Lord, have mercy.

DEACON. That the whole day may be perfect, holy, peaceful and sinless, we ask the Lord.

CATHEDRAL CHOIR [*singing*]. Vouchsafe, O Lord.

DEACON. An angel of peace, a true teacher, a saviour of our souls and bodies, we ask the Lord.

CHOIR [*sings*]. Vouchsafe, O Lord.

VOICES. 'The deacon looks as if he's going on for ever with his "Have mercy, O Lords" and his "Vouchsafe, O Lords".' 'I'm fed up with standing about.'

DEACON. Pardon and forgiveness of our sins and transgressions we ask the Lord.

CHOIR [*sings*]. Vouchsafe, O Lord.

DEACON. What is good and profitable for our souls, and peace unto the world, we ask the Lord.

A VOICE. 'They're being rowdy again. What awful people!'

CHOIR [*sings*]. Vouchsafe, O Lord.

VERA. Peter, you're shuddering and breathing heavily. Are you going to faint?

SABININ. The woman in black, she—it's our fault.

VERA. What woman?

[*A groan.*]

SABININ. It's Tatyana groaning. I'm trying to pull myself together. Kotelnikov's pressing the crown on my head. Oh, all right—never mind.

DEACON. That the remaining time of our life may end in peace and repentance, we ask the Lord.

CHOIR. Vouchsafe, O Lord.

KOKOSHKIN. Vera's as white as a sheet. Look, there are tears in her eyes, I think. And he—look at him!

MRS. KOKOSHKIN. I told her that people would behave badly. I can't think why she decided to be married here—why didn't she go to a village church?

DEACON. A christian end of our life without sickness or shame, in peace and able to answer at the last judgement of Christ, we ask it.

CHOIR [*sings*]. Vouchsafe, O Lord.

MRS. KOKOSHKIN. We ought to ask Father Ivan to hurry. She looks like death.

VOLGIN. I say, let me take over. [*Takes* KOTELNIKOV'*s place.*]

DEACON. Having prayed for the unity of the faith, and the communion of the Holy Ghost both for ourselves and for each other, let us make over our whole lives to Christ our God.

CHOIR [*sings*]. To Thee, O Lord.

SABININ. Pull yourself together, Vera—be like me. Ah, well. The service will soon be over anyway. We'll leave at once. It's her——

VOLGIN. Shush!

FATHER IVAN. And count us worthy, O Master, with boldness to dare without condemnation to call upon thee, our heavenly Father God, and say——

BISHOP'S CHOIR [*sings*]. Our Father which art in heaven, hallowed be thy name, thy kingdom come——

MATVEYEV [*to the actors*]. Make way a bit, boys, I want to kneel. [*Kneels and bows to the ground.*] Thy will be done, in heaven as on earth. Give us this day our daily bread and forgive us our sins——

BISHOP'S CHOIR. Thy will be done, as in heaven—in heaven—our daily bread—daily——

MATVEYEV. Remember, O Lord, thy deceased handmaiden TATYANA and forgive her her sins of commission and omission, and forgive us and have mercy on us—. [*Stands up.*] It's hot!

BISHOP'S CHOIR. —give us this day and forgive us—and forgive us our trespasses—as we forgive them that trespass against us—us——

VOICE. 'They *are* dragging the thing out, I must say.'

BISHOP'S CHOIR. —and lead us—us not into temptation, but deliver us from e-e-evil.

KOTELNIKOV [*to the* ASSISTANT PROSECUTOR]. What's bitten the groom? See him trembling?

ASSISTANT PROSECUTOR. What's the matter with him?

KOTELNIKOV. That woman in black who just had hysterics—he thought she was Tatyana. He must be seeing things.

FATHER IVAN. For thine is the kingdom, the power and the glory, Father, Son and Holy Ghost, now and forever, world without end.

CHOIR. Amen.

ASSISTANT PROSECUTOR. Mind he doesn't do something silly.

KOTELNIKOV. He won't break down—not him!

ASSISTANT PROSECUTOR. Yes, it's pretty tough on him.

FATHER IVAN. Peace to all.

CHOIR. And on thy spirit.

DEACON. Bow your heads to the Lord.

CHOIR. To thee, O Lord.

VOICES. 'I think they're going to have the procession now. Shush!' 'Have they had an inquest on the doctor's wife?' 'Not yet. They say her husband left her. And Sabinin abandoned Tatyana, didn't he? That's true, isn't it?' 'Ye-es.' 'I remember the inquest on Miss Repin.'

DEACON. Let us pray to the Lord.

CHOIR. Lord, have mercy.

FATHER IVAN [*reads*]. O God, who by thy might createst all things, and confirmest the universe, and adornest the crown of all things created by thee; do thou, with thy spiritual blessing, bless also this common cup given for the community of marriage unto them that are conjoined. For blessed is thy name, and glorified thy kingdom, Father, Son and Holy Ghost, now and forever, world without end. [*Gives* SABININ *and* VERA *wine to drink.*]

CHOIR. Amen.

ASSISTANT PROSECUTOR. Mind he doesn't faint.

KOTELNIKOV. He's a rugged brute, he won't break down.

VOICES. 'Stick around, all of you—we'll all go out together. Is Zipunov here?' 'Yes. We must stand round the carriage and hiss for five minutes or so.'

FATHER IVAN. Give me your hands, please. [*Ties* SABININ'*s and* VERA'*s hands together with a handkerchief.*] Not too tight?

ASSISTANT PROSECUTOR [*to the* STUDENT]. Give me the crown, boy, and you carry the train.

BISHOP'S CHOIR [*sings*]. Rejoice, O Esaias, the virgin is with child——

[FATHER IVAN *walks round the lectern followed by the bride and groom and their attendants.*]

VOICE. 'The student's got tangled up in the train.'

BISHOP'S CHOIR. —and bringeth forth a son, Emmanuel, God and man: the orient is his name——

SABININ [*to* VOLGIN]. Is that the end?

VOLGIN. Not yet.

BISHOP'S CHOIR. —whom magnifying, we call the virgin blessed.

[FATHER IVAN *walks around the lectern for the second time.*]

CATHEDRAL CHOIR [*sings*]. O holy martyrs, who valiantly contended, and are crowned; pray ye the Lord for mercy on our souls——

FATHER IVAN [*goes round for the third time and sings*]. —on our souls.

SABININ. My God, will this never end?

BISHOP'S CHOIR [*sings*]. Glory to thee, Christ God, apostles' boast and martyrs' joy, whose preaching was the consubstantial Trinity.

OFFICER IN THE CONGREGATION [*to* KOTELNIKOV]. Warn Sabinin that the students and grammar school boys are going to hiss him in the street.

KOTELNIKOV. Thank you. [*To the* ASSISTANT PROSECUTOR.] But how long this business drags out! Will the service never be done? [*Wipes his face on his handkerchief.*]

ASSISTANT PROSECUTOR. But your own hands are shaking—you're such a lot of sissies.

KOTELNIKOV. I can't get Tatyana out of my head. I keep imagining Sabinin singing and her crying.

FATHER IVAN [*taking the crown from* VOLGIN, *to* SABININ]. Be thou magnified, O bridegroom, as Abraham, and blessed as Isaac, and increased as Jacob, walking in peace, and performing in righteousness the commandments of God.

YOUNG ACTOR. What pearls to cast before swine like that!

MATVEYEV. It's the same God for us all.

FATHER IVAN [*taking the crown from the* ASSISTANT PROSECUTOR, *to* VERA]. And thou, O bride, be thou magnified as Sara, and rejoiced as Rebecca, and increased as Rachel, being glad in thy husband, and keeping the paths of the law, for so God is well pleased.

[*There is a great rush towards the exit.*]

VOICES. 'Be quiet, all of you. It's not over yet.' 'Shush! Don't push!'

DEACON. Let us pray to the Lord.

CHOIR. Lord, have mercy.

FATHER ALEXIS [*reads, after taking off his dark glasses*]. O God, our God, who wast present in Cana of Galilee, and didst bless the marriage there; do thou bless also these thy servants, who, by thy providence, are conjoined in the community of marriage. Bless their incomings and outgoings, replenish their life with good things, accept their crowns in thy kingdom unsullied and undefiled, and preserve them without offence to ages of ages.

CHOIR [*sings*]. Amen.

VERA [*to her brother*]. Ask them to fetch me a chair. I feel ill.

STUDENT. It will soon be over. [*To the* ASSISTANT PROSECUTOR.] Vera feels ill.

ASSISTANT PROSECUTOR. It'll be over in a moment, Vera. Just a moment. Hold on, my dear.

VERA [*to her brother*]. Peter doesn't hear me, he's like a man in a trance. My God, my God! [*To* SABININ.] Peter!

FATHER IVAN. Peace to all.

CHOIR. And on thy spirit.

DEACON. Bow your heads to the Lord.

FATHER IVAN [*to* SABININ *and* VERA]. The Father, the Son, and the Holy Ghost, the all-holy, and consubstantial and life-originating Trinity, one Godhead and sovereignty, bless you, and vouchsafe unto you long life, well-favoured children, progress in life and faith, and replenish you with all the good things of earth, and count you worthy of the obtaining of promised blessings, through the prayers of the holy God-bearing one, and of all the Saints. Amen! [*To* VERA, *with a smile.*] Kiss your husband.

VOLGIN [*to* SABININ]. Don't just stand around—kiss each other.

[*The bride and groom kiss.*]

FATHER IVAN. Congratulations. God grant——

MRS. KOKOSHKIN [*goes up to* VERA]. My dear, I'm so glad, darling. Congratulations!

KOTELNIKOV [*to* SABININ]. Congratulations on getting spliced. No need to look pale any more, the whole rigmarole's over.

DEACON. Wisdom!

[*Congratulations.*]

CHOIR [*sings*]. Thee more honourable than the Cherubim, and incomparably more glorious than the Seraphim, thee who bearest without corruption God the word, O true Mother of God, thee we magnify! In the Lord's name, bless us, Father.

[*The crowd presses out of the church.* KUZMA *puts out the large candles.*]

FATHER IVAN. May Christ, our true God, that by his presence in Cana of Galilee declared marriage to be honourable, Christ our true God, through the prayers of his most pure Mother; of the holy, glorious, and all-praised apostles; of the holy god-crowned sovereigns and equals of the apostles, Constantine and Helen; of the holy great martyr Procopius, and of all the Saints, have mercy upon us and save us, as being good and the lover of mankind.

CHOIR [*singing*]. Amen. Lord, have mercy. Lord, have mercy. Lord, have me-e-rcy.

LADIES [*to* VERA]. Congratulations, dear. May you be very happy! [*Kisses.*]

SONNENSTEIN. Mrs. Sabinin, er, to put it in plain language, isn't it——

BISHOP'S CHOIR [*singing*]. Long life, long life, lo-o-ng life!

SABININ. So sorry, Vera. [*Grips* KOTELNIKOV'*s arm and quickly takes him to one side; trembling and choking.*] Let's go to the cemetery immediately.

KOTELNIKOV. You must be out of your mind. It's night now. What are you going to do there?

SABININ. Come on, for God's sake—please.

KOTELNIKOV. Take your bride home, you lunatic.

SABININ. I don't give a damn, blast everything to hell! I—I'm going. Must hold a requiem service. But I must be mad—I almost died. Oh, Kotelnikov, Kotelnikov!

KOTELNIKOV. Come, come.

[*Leads him to his bride. A minute later a piercing whistle can be heard from the street. The congregation gradually leaves the church. Only the* ACOLYTE *and* KUZMA *remain behind.*]

KUZMA [*puts out the candelabras*]. What a huge crowd——

ACOLYTE. Well, yes. It's a smart wedding. [*Puts on a fur coat.*] These people know how to live.

KUZMA. It's all so pointless. No sense in it.

ACOLYTE. What?

KUZMA. Take this wedding. We marry them, christen them and bury them every day and there ain't no sense in it all.

ACOLYTE. Then what exactly would you like to do?

KUZMA. Nothing. Nothing at all. What's the point? They sing, burn incense, recite the liturgy—but God don't listen. Forty years I've worked here and never heard God's voice. Where God is, I just don't know. There's no point in anything.

ACOLYTE. Well, yes. [*Puts on his galoshes.*] 'It's talk like this that makes your head go round.' [*Moves off with squeaking galoshes.*] Cheerio! [*Goes out.*]

KUZMA [*on his own*]. We buried a local squire this afternoon, we've just had this wedding, and we've a christening tomorrow morning. Where will it all end? What use is it to anyone? None, it's just pointless.

[*A groan is heard.* FATHER IVAN *and* FATHER ALEXIS, *he of the flowing locks and dark glasses, come out from the chancel.*]

FATHER IVAN. I daresay he picked up a decent dowry too.

FATHER ALEXIS. Bound to have.

FATHER IVAN. Oh, what a life, when you come to look at it. Why, I courted a girl myself once, married her and received a dowry, but that's all buried in the sands of time now. [*Shouts.*] Kuzma, why did you put out all the candles? You'll have me falling over in the dark.

KUZMA. Oh, I thought you'd left.

FATHER IVAN. Well, Father Alexis? Shall we go and have tea over at my place?

FATHER ALEXIS. No, thank you, Father. I've no time, there's another report to write.

FATHER IVAN. As you wish.

WOMAN IN BLACK [*comes out from behind a column, staggering*] Who's there? Take me away, take me away.

FATHER IVAN. What's this? Who is it? [*Frightened.*] What do you want, my dear?

FATHER ALEXIS. Lord forgive us miserable sinners.

WOMAN IN BLACK. Take me out, please. [*Groans.*] I'm the sister of Ivanov, the army officer—his sister.

FATHER IVAN. What are you doing here?

WOMAN IN BLACK. I've taken poison—because I hate him. He insulted—. So why is he so happy? My God! [*Shouts.*] Save me, save me! [*Sinks to the floor.*] Everyone should take poison, everyone! There's no justice in this world!

FATHER ALEXIS [*in horror*]. What blasphemy! O God, what blasphemy!

WOMAN IN BLACK. Because I hate him. We should all poison ourselves. [*Groans and rolls on the floor.*] She's in her grave and he, he—the wrong done to a woman is a wrong done to God. A woman has been destroyed.

FATHER ALEXIS. What blasphemy against religion! [*Throws up his arms.*] What blasphemy against life!

WOMAN IN BLACK [*tears all her clothes and shouts*]. Save me! Save me! Save me!

CURTAIN

And all the rest I leave to A. S. Suvorin's imagination.

A TRAGIC ROLE

(A holiday episode)

[*Трагик по неволе (из дачной жизни)*]

A FARCE IN ONE ACT

(1889–1890)

CHARACTERS

IVAN TOLKACHOV, a family man

ALEXIS MURASHKIN, his friend

The action takes place in Murashkin's flat in St. Petersburg

MURASHKIN'*s study. Easy chairs.* MURASHKIN *is sitting at his desk.* TOLKACHOV *comes in. He carries a glass lamp-globe, a child's bicycle, three hat-boxes, a large bundle of clothes, a shopping-bag full of beer and a lot of small parcels. He rolls his eyes in a dazed sort of way, and sinks exhausted on the sofa.*

MURASHKIN. Hallo, old man—nice to see you! Where are you sprung from?

TOLKACHOV [*breathing hard*]. Will you do me a favour, dear boy? Have a heart—lend me a revolver till tomorrow, there's a good chap.

MURASHKIN. A revolver? Whatever for?

TOLKACHOV. I need one. Oh Lord, give me water! Water, quick! I need one—got to drive through a dark wood tonight, so I, er—always be prepared. Lend me one, please.

MURASHKIN. Oh, rubbish, old man! 'Dark wood'—what the devil do you mean? Up to something, aren't you? And up to no good too, from the look of you. Now, what's the matter? Are you ill?

TOLKACHOV. Stop, let me get my breath. Whew, I'm all in! My whole body, the old brain-box—I feel as if they'd been chopped up and stuck on a skewer. Like a dog's dinner I feel, I'm at the end of my tether. Don't ask questions or go into details, there's a good fellow—just give me the revolver, I implore you.

MURASHKIN. Come, come, old boy, don't say you've got the wind up! And you a family man, high up in the civil service and all that! You should be ashamed!

TOLKACHOV. A family man? A martyr, more like—a complete drudge, a slave, a chattel, the lowest thing that crawls. Why don't I end it all—what am I waiting for, like some benighted idiot? I'm a regular doormat. What have I to live for? Eh? [*Jumps up.*] Come on, tell me what I have to live for, why the unbroken chain of moral and physical tortures? Certainly I can understand a man who sacrifices himself for an ideal, but to martyr yourself to women's petticoats, lampshades and that sort of damn tomfoolery—no, thank you very much. No, no, no! I've had enough! Enough, I tell you!

MURASHKIN. Don't talk so loud, the neighbours will hear.

TOLKACHOV. Then let the neighbours hear, I don't care. If you won't lend me a revolver, someone else will—I'm not long for this world anyway, that's settled.

MURASHKIN. Hey, you've pulled off one of my buttons. Cool off, can't you? What's wrong with your life? That's what I still don't see.

TOLKACHOV. Oh, don't you! What's wrong, he asks! All right then, I'll tell you. By all means! I'll get it off my chest and perhaps I'll feel better. Let's sit down. Now, you listen—. Lord, I'm out of breath. Take today, for instance. Yes, take it by all means! From ten till four I have my nose to the old office grindstone, as well you know. There's this ghastly heat-wave, the flies—sheer chaos, old boy, all hell let loose. My secretary's on leave, my assistant's gone off to get married, the office rank and file are all mad on week-ends in the country, love affairs and play-acting. The whole bunch are so drowsy, jaded and haggard, you can get no sense out of them. The secretary's job's being done by a creature who's deaf in his left ear and in love. Our clients are crazy, all in a tearing hurry, all losing their tempers and threatening us—there's such a god-almighty hullabaloo, it's enough to make you scream. It's a real mix-up—sheer hell on earth. And the work itself is so damn awful, the same thing over and over again, answering endless inquiries, writing endless minutes—just one damn boring thing after another. It's enough to send you pop-eyed, see? Give me water—. You come out of the office dead beat, absolutely whacked. All you're fit for is a meal and flopping into bed, but not a bit of it! Don't forget your family's staying out at the holiday cottage—in other words you're a complete dogsbody, a worm, a nonentity, an errand-boy expected to rush round like a scalded cat at everyone's beck and call. There's a charming custom where we're staying: if a husband travels in to the office every day, then not only his wife, but any little squirt in the place is fully entitled to load him with masses of errands. The wife tells me to go to her dressmaker and kick up a row because she made a bodice too full, but too narrow in the shoulders. Sonya needs a pair of shoes changed, my sister-in-law wants twenty copecks' worth of crimson silk to match a pattern, and seven foot of ribbon. Just a moment—here, I'll read it to you. [*Takes a list out of his pocket and reads.*] One globe for lamp; one pound ham-sausage; five copecks' worth of cloves and cinnamon; castor oil for Misha; ten pounds granulated sugar; fetch copper bowl and mortar for pounding sugar from home; carbolic acid; insect powder; ten copecks' worth face powder; twenty bottles beer; vinegar and a size eighty-two corset for Mlle Chanceau—ugh! And fetch Misha's overcoat and galoshes from home. So much for the domestic order. Now for my charming friends and neighbours, blast 'em—now for *their* commissions. It's young Vlasin's birthday tomorrow, so I have to buy him a bicycle. Colonel Vikhrin's missus is in the family way, so it's call on the midwife every day and ask her to come along. And so on and so forth. I've five lists in my pocket and my handkerchief's all knots. And so, old man, between leaving the office and catching your train you dash round town like a mad dog with your tongue hanging out, rushing here there and everywhere and cursing your fate. You traipse around from draper's to chemist's, from chemist's to dressmaker's, from dressmaker's to sausage-shop, then back to the chemist's again. You trip over yourself, you lose your money, you forget to pay, so they run after you and kick up a fuss, and somewhere else you tread on a woman's skirt—ugh! This mad rush puts you in a frenzy, makes you ache something chronic all night through, and you dream of crocodiles. All right, your errands are done, everything's bought—but how exactly are you going to parcel up all this junk? Are you going to lump the lampshade along with your heavy copper mortar and pestle, for instance? Or mix carbolic acid and tea? How do you fit the beer bottles in with this bicycle? It's a regular labour of Hercules, it is—a proper Chinese puzzle. Rack your brains, flog your wits as you like, you always end up breaking and spilling something, then on the station and in the train you find yourself standing with your arms stuck out and your legs straddled, a package under your chin, and with shopping-bags, cardboard boxes and the rest of the tripe strung all over you. Then the train starts and the passengers chuck your stuff all over the place because you've put it on people's seats. They shout, call the ticket-collector, say they'll have you put off. And all you can do is stand there cringing like a whipped cur. And then what? You reach the cottage keyed up for a real drink after all these good works—now for a meal and a bit of a snooze, eh? Some hope! The wife's had her eye on you all along, and you've scarce downed a mouthful of soup, poor devil, before she shoves her dirty great oar in: 'You'd like to see the amateur theatricals, wouldn't you, dear? Or go to the dancing club?' Just you try and get out of it! You're a husband. And in holiday parlance the word 'husband' means a dumb beast to be driven and loaded with all the baggage in sight—no

Society for the Prevention of Cruelty to Animals to bother about! You go and goggle at *A Scandal in a Respectable Family* or some other damn silly play, you clap when the wife tells you, you feel your strength ebbing away, and expect to have an apoplectic fit any minute. And at the club you've got to watch the dances and find partners for the wife, and if there aren't enough gentlemen to go round you have to dance the quadrille yourself. You get home from that theatre or dance after midnight more dead than alive, and fit for nothing. But you've made it at last, and it's off with your various trappings and climb into bed! Wonderful! Just shut your eyes and sleep. Nice and snug, isn't it—like a dream come true? The children aren't screaming in the next room, the wife isn't there, and you've nothing on your conscience—what could be better? You drop off. Then, all of a sudden—bzzzzzz! Gnats! [*Jumps up.*] Gnats! Damn and blast those bleeding gnats! [*Shakes his fist.*] Gnats! The Plagues of Egypt aren't in it—*or* the Spanish Inquisition! Buzz, buzz, buzz! There's this pathetic, mournful buzzing as if it was saying how sorry it was, but you wait till the little nipper gets his fangs in—it means an hour's hard scratching. You smoke, you lash out, you shove your head under the blankets—but you're trapped. You end by giving yourself up like a lamb to the slaughter—let the bastards have their meal and get stuffed! But before you've got used to the gnats, horror strikes again! Your wife starts practising songs in the drawing-room with some friends. Tenors! They sleep all day and spend their nights getting up amateur concerts. Ye gods! For sheer undiluted hell give me your average tenor! Gnats aren't in the same street! [*Sings.*] 'Oh, tell me not your young life's ruined.' 'Once more I stand entranced before thee.' It's the death by a thousand cuts, blast 'em! To deaden their voices a bit, I have this dodge of tapping my head near the ear. So there I am tapping away till they leave at about four o'clock. Whew! More water, old boy, I'm about played out. Well, anyway, at six o'clock you get up after a sleepless night and set off to the station to catch your train. You dash along afraid of missing it. The mud! The mist! And the cold—brrr! And when you get to town the whole ruddy rigmarole starts all over again. That's the way of it, old boy. It's a rotten life, I can tell you—I wouldn't wish it on my worst enemy. It's made me ill, you know—I've asthma, I've heartburn, I'm always on edge, I've indigestion and these dizzy spells. I've become quite a psychopath, you know! [*Looks about him.*] Keep this under your hat, but I feel like calling in one of our leading head-shrinkers. I get this funny mood, old man. You see, when you're really bitched, bothered and bewildered, when gnats do bite or tenors do sing, suddenly everything goes blank and you jump up and run round the house as if you were berserk, shouting: 'I must have blood! Blood!' At such times you really do feel like sticking a knife in someone's gizzard or bashing his head in with a chair. That's what this holiday commuting does for you! And you get no sympathy or pity, either, everyone takes it so much for granted. You even get laughed at. But I'm alive, aren't I? So I want a bit of life! This isn't funny, it's downright tragic. Look here, if you won't lend me a revolver, at least show a spot of fellow-feeling.

MURASHKIN. But I do feel for you.

TOLKACHOV. Yes, I can see how much you do. Well, good-bye. I'll get some sprats and salami, er, and I want some toothpaste too—and then to the station.

MURASHKIN. Whereabouts are you taking your holiday?

TOLKACHOV. At Corpse Creek.

MURASHKIN [*joyfully*]. Really? I say, you don't happen to know someone staying there called Olga Finberg?

TOLKACHOV. Yes. In fact she's a friend of ours.

MURASHKIN. You don't say! Well, I never! What a stroke of luck and how nice if you——

TOLKACHOV. Why, what is it?

MURASHKIN. My dear old boy, could you possibly do me a small favour, there's a good chap? Now, promise me you will.

TOLKACHOV. What is it?

MURASHKIN. Be a friend in need, old man—have a heart! Now, first give my regards to Olga and say I'm alive and well, and that I kiss her hand. And secondly, there's a little thing I want you to take her. She asked me to buy her a sewing-machine and there's no one to deliver it. You take it, old man. And you may as well take this cage with the canary while you're about it—only be careful or the door will break. What are you staring at?

TOLKACHOV. Sewing-machine—. Cage—. Canary—. Why not a whole bloody aviary?

MURASHKIN. What's up, man? Why so red in the face?

TOLKACHOV [*standing*]. Give me that sewing-machine! Where's the bird-cage? Now you get on my back too! Gobble me up! Rend me limb from limb! Do me in! [*Clenching his fists.*] I must have blood! Blood!

MURASHKIN. Have you gone mad?

TOLKACHOV [*bearing down on him*]. I must have blood! Blood! Blood!

MURASHKIN [*in horror*]. He's gone mad! [*Shouts.*] Peter! Mary! Where are the servants? Help!

TOLKACHOV [*chasing him round the room*]. I must have blood! Blood, blood, blood!

CURTAIN

THE WEDDING

[*Свадьба*]

A PLAY IN ONE ACT

(1889–1890)

CHARACTERS

YEVDOKIM ZHIGALOV, a minor civil servant, retired

NASTASYA, his wife

DASHA, their daughter

EPAMINONDAS APLOMBOV, her fiancé

COMMANDER THEODORE REVUNOV-KARAULOV, Imperial Russian Navy, retired

ANDREW NYUNIN, insurance agent

MRS. ANNA ZMEYUKIN, midwife; wears a bright crimson dress; aged 30

IVAN YAT, telegraph clerk

KHARLAMPY DYMBA, a Greek confectioner

DMITRY MOZGOVOY, sailor in the Volunteer Fleet

The best man, other young men, servants etc.

The action takes place in a reception room at a second-class restaurant

A large, brilliantly lit room. A big table laid for supper with tail-coated waiters fussing round it. A band plays the last figure of a quadrille off-stage.

MRS. ZMEYUKIN, YAT *and the* BEST MAN *cross the stage.*

MRS. ZMEYUKIN. No, no, no!

YAT [*following her*]. Oh come on, have a heart.

MRS. ZMEYUKIN. No, no, no!

BEST MAN [*hurrying after them*]. This won't do, you two. Where do you think you're off to? What about the dancing? *Grand-rond, seel voo play!*

[*They go out.* MRS. ZHIGALOV *and* APLOMBOV *come in.*]

MRS. ZHIGALOV. Don't bother me with this stuff—you go and dance instead.

APLOMBOV. I'm not spinning round like a top, thank you. Me spin? No sir, I'm no Spinoza! I'm a solid citizen, a pillar of society, and I am not amused by such idle pursuits. But the dancing's neither here nor there. I'm sorry, Mother, but some things you do have me baffled. Take the dowry, for instance. Besides certain household utilities, you also promised me two lottery tickets. Well, where are they?

MRS. ZHIGALOV. I've got a bit of a headache. It must be the weather, there's a thaw in the offing.

APLOMBOV. Don't you try and fob me off! I've just found out that you've pawned those tickets. I'm sorry, Mother, but that's a pretty mean trick. I don't say so from selfishness—I don't want your lottery tickets, it's the principle of the thing. No one makes a monkey out of me. I've brought your daughter happiness, but if I don't get those tickets today I'll make her wish she'd never been born—as I'm a man of honour!

MRS. ZHIGALOV [*looking at the table and counting the number of places laid*]. One, two, three, four, five——

A WAITER. The chef asks how you want the ice-cream served—with rum or madeira or on its own?

APLOMBOV. Rum. And you can tell your boss there's not enough wine, Get him to bring on more Sauterne. [*To* MRS. ZHIGALOV.] You also promised to invite a General to this evening's celebration—that was clearly understood. Well, where is he, I should like to know?

MRS. ZHIGALOV. It's not my fault, dear.

APLOMBOV. Whose is it then?

MRS. ZHIGALOV. Andrew Nyunin's. He was here yesterday and promised to bring along a General—the genuine, guaranteed article. [*Sighs.*] He can't have run across any, or he'd certainly have brought one. It's not that we're mean—nothing's too much for our darling daughter. If you want a General, a General you shall have.

APLOMBOV. And another thing, Mother. Everyone, you included, knows that this telegraph-clerk Yat was going round with Dasha before I asked her to marry me. Why did you have to invite him? Surely you knew I'd find it awkward?

MRS. ZHIGALOV. Oh—what's your name?—Epaminondas, you haven't been married twenty-four hours and you've nearly been the death of me and Dasha already with your talk, talk, talk. What will it be like after a year of it? You do nag so, you really do.

APLOMBOV. So you don't like hearing a few home truths, eh? Just as I thought. Then behave yourself. That's all I ask, behave properly.

[*Couples dance the* grand-rond, *crossing the room from one door to the other. The first couple are* DASHA *and the* BEST MAN, *the last consists of* YAT *and* MRS. ZMEYUKIN. *These last two drop behind and stay in the ballroom.* ZHIGALOV *and* DYMBA *come in and go up to the table.*]

BEST MAN [*shouts*]. *Promenade! Mess-sewers, promenade!* [*Off-stage.*] *Promenade!*

[*The couples go out.*]

YAT [*to* MRS. ZMEYUKIN]. Have a heart—take pity on us, enchanting Anna Zmeyukin.

MRS. ZMEYUKIN. Oh, you are a one, really! I'm not in voice tonight, I've told you that already.

YAT. Sing something, I beg you, if only one note! For pity's sake! A single note!

MRS. ZMEYUKIN. Don't be such a bore. [*Sits down and waves her fan.*]

YAT. Oh, you're quite heartless! That a being so cruel, pardon the expression, should have such a simply divine voice! You shouldn't be a midwife with a voice like that, if you'll pardon my saying so—you should sing in the concert hall. For instance, there's the heavenly way you take that twiddly bit—how does it go? [*Sings softly.*] 'I loved you, but my love in vain—.' Superb!

MRS. ZMEYUKIN [*sings softly*]. 'I loved you once and love perhaps might still—.' Is that it?

YAT. That's the one. Super!

MRS. ZMEYUKIN. No, I'm not in voice today. Here—you fan me, it's hot! [*To* APLOMBOV.] Why so downhearted, Mr. Aplombov? And on your wedding day too! You should be ashamed of yourself, you naughty man. Well, why so pensive?

APLOMBOV. Marriage is a serious step. You have to weigh the whole thing up from every angle.

MRS. ZMEYUKIN. What revolting cynics you all are. This atmosphere makes me choke. Give me air, do you hear? Atmosphere! [*Sings softly.*]

YAT. Too, too divine!

MRS. ZMEYUKIN. Fan me, fan me! Or I think I'll burst. Tell me, please, why do I have this choking feeling?

YAT. Because you're sweating.

MRS. ZMEYUKIN. Oh, don't you be so common! How dare you speak to me like that?

YAT. Pardon, I'm sure. Of course, you're used to high society, if you don't mind my saying so, and——

MRS. ZMEYUKIN. Oh, let me alone. Give me romance, excitement! Fan me, fan me!

ZHIGALOV [*to* DYMBA]. Have another? [*Pours.*] A drink always comes in handy. The great thing is—don't neglect your business, my dear Dymba. Drink, but keep your wits about you. And if you want a little drink, why not have a little drink? A little drink does no harm. Your health! [*They drink.*] Tell me, are there tigers in Greece?

DYMBA. Yes, is tigers.

ZHIGALOV. And lions?

DYMBA. Is lions too. In Russia is nothing, in Greece is everything! In Greece is my father, my uncle, my brothers. Here is nothing, isn't it?

ZHIGALOV. I see. And are there whales in Greece?

DYMBA. In Greece is every damn thing.

MRS. ZHIGALOV [*to her husband*]. Why are you all eating and drinking any old how? It's time everyone sat down to table. Don't stick your fork in the lobsters, they're meant for the General. Perhaps he'll still come——

ZHIGALOV. Have you got lobsters in Greece?

DYMBA. We have. In Greece is damn all, I tell you!

ZHIGALOV. I see. Have you established civil servants too?

MRS. ZMEYUKIN. I can imagine what a terrific atmosphere there is in Greece!

ZHIGALOV. And a terrific lot of funny business goes on too, I'll bet. The Greeks are just like the Armenians or gipsies, aren't they? Can't sell you a sponge or a goldfish without trying to do you down. How about another?

MRS. ZHIGALOV. Why keep knocking it back? It's time everyone sat down, it's nearly midnight.

ZHIGALOV. If we're to sit, then sit we will. Ladies and gentlemen, I humbly beg you—this way, please. [*Shouts.*] Supper's ready! Come on, young people.

MRS. ZHIGALOV. Come and have supper, please, all of you. Take your places.

MRS. ZMEYUKIN [*sitting down at table*]. Give me a bit of poetry!

'Restless, he seeks the raging storm,
As if the storm could give him rest.'

YAT [*aside*]. Superb creature! Oh for a storm! I'm in love, head over heels!

[*Enter* DASHA, MOZGOVOY, BEST MAN, *young men, girls and so on. All sit down noisily at table. There is a minute's pause. The band plays a march.*]

MOZGOVOY [*standing up*]. Ladies and gentlemen, I've something to say. There are lots of toasts and speeches to come, so let's not mess about, but plunge straight in. Ladies and gentlemen, I propose: the bride and bridegroom!

[*The band plays a flourish. Cheers. Clinking of glasses.*]

MOZGOVOY. The bride and bridegroom!

ALL. The bride and bridegroom!

[APLOMBOV *and* DASHA *kiss.*]

YAT. Superb, divine! I must say, ladies and gentlemen—always give credit where credit is due—this room and whole establishment are magnificent! They're terrific, charming! But you know, we do need one thing to set things off to perfection—electric light, if you'll pardon the expression. Electric light's come in all over the world—only Russia lags behind.

ZHIGALOV [*with an air of profundity*]. Electric light. I see. If you ask me, there's a lot of funny business about electric light. They shove in a little bit of coal and think no one will notice. No, dear boy, if we're to have light, don't give us your coal, but something with a bit of body to it, something solid that a man can get his teeth in. Give us real light, see? Natural light, not something imaginary.

YAT. If you'd seen what an electric battery's made of, you'd tell a different tale.

ZHIGALOV. I don't want to see. It's all a lot of funny business, to cheat the common man—squeeze him dry, they do, we know their sort! As for you, young feller-me-lad, don't you stick up for swindlers. Have a drink instead and fill up the glasses, that's my message to you!

APLOMBOV. I quite agree, Dad old man. Why trot out all the long words? Not that I mind discussing modern inventions, like, in a scientific manner of speaking. But there's a time and place for everything. [*To* DASHA.] What do you think, dear?

DASHA. The gentleman's only trying to show how brainy he is, talking about things no one understands.

MRS. ZHIGALOV. We've lived our lives without book-learning, praise the Lord, and it didn't stop us finding good husbands for three daughters. If you think us so ignorant, why come here? Go and visit your smart friends.

YAT. I've always thought highly of your family, Mrs. Zhigalov, and if I did mention electric light, it doesn't mean I was trying to show off. I'm quite prepared to have a drink. I've always wished Dasha a good husband with all my heart. They don't grow on trees these days, Mrs. Zhigalov, good husbands don't. Nowadays everyone's out for what he can get, they all want to marry for money.

APLOMBOV. That's an insinuation!

YAT [*taking fright*]. No harm intended, I'm sure. I wasn't speaking of present company, it was just, er, a general remark. Oh, for heaven's sake—everyone knows you're marrying for love. It's not as if the dowry was up to much!

MRS. ZHIGALOV. Oh, isn't it? You mind your Ps and Qs, young man. Besides a thousand roubles in cash, we're giving three lady's coats, a bed and all the furniture. You'll not find many dowries to match that!

YAT. I meant no harm. Certainly the furniture's nice, and, er, so are the coats, of course. I was merely concerned with this gent being offended on account of my insinuations.

MRS. ZHIGALOV. Then don't *make* any. We ask you to the wedding out of regard for your mother and father, and now we get all these remarks! If you knew Mr. Aplombov was marrying for money, why not say so before? [*Tearfully*.] I've reared her, nurtured her, looked after her. She was the apple of her mother's eye, my darling little girl——

APLOMBOV. You mean you actually believe him! Thank you very much! Most grateful, I'm sure! [*To* YAT.] Mr. Yat, though you're a friend of mine, I won't have you behaving so outrageously in other people's houses. Be so good as to make yourself scarce!

YAT. I *beg* your pardon!

APLOMBOV. I wish you were as much of a gentleman as what I am. In a word, kindly buzz off!

[*The band plays a flourish.*]

YOUNG MEN [*to* APLOMBOV]. Oh, leave him alone, can't you? Stop it. What's the good? Sit down. Leave him alone.

YAT. I never said a thing, I was only—. I must say, I can't see why—. All right, I'll go then. But first you pay back the five roubles you borrowed last year to buy yourself a fancy, pardon the expression, waistcoat. I'll have another drink and, er, go. But you pay up first.

YOUNG MEN. Oh, stop it, stop it! That'll do. A lot of fuss about nothing.

BEST MAN [*shouts*]. To the health of the bride's mother and father! Mr. and Mrs. Zhigalov!

[*The band plays a flourish. Cheers.*]

ZHIGALOV [*bows in all directions, deeply moved*]. Thank you very much. My dear guests, I'm most grateful to you for remembering us and coming along and not turning up your noses. Now, don't go thinking this is all a put-up job or a lot of funny business. I'm just saying what I feel, speaking from the bottom of my heart. Nothing's too good for decent folk. My humble thanks. [*Exchanges kisses.*]

DASHA [*to her mother*]. Why are you crying, Mum? I'm so happy.

APLOMBOV. Your mother's upset at the thought of being separated from you. My advice to her is—remember what I told her just now.

YAT. Don't cry, Mrs. Zhigalov. Think what human tears are—a weakness in the psychology department, that's all.

ZHIGALOV. Are there mushrooms in Greece?

DYMBA. Is plenty. Is everything.

ZHIGALOV. Well, I bet there aren't any white and yellow ones like ours.

DYMBA. Is white. Is yellow. Is everything.

MOZGOVOY. Mr. Dymba, it's your turn to make a speech. Let him speak, ladies and gentlemen.

ALL [*to* DYMBA]. Speech! Speech! Your turn!

DYMBA. Pliss? Do not understand—. What is which?

MRS. ZMEYUKIN. Oh, no you don't—don't you dare try and wriggle out of it! It's your turn. Up you get!

DYMBA [*stands up in embarrassment*]. I speak, isn't it? Is Russia. Is Greece. Russian peoples is in Russia. Greek peoples is in Greece, isn't it? In sea is sailing-boots that Russians call sheep. Is railways on land. I am understanding very well, isn't it? We are Greeks, you are Russians and I am not needing anything, isn't it? I am also saying—. Is Russia. Is Greece——.

[NYUNIN *comes in.*]

NYUNIN. Just a moment, all of you—don't start eating just yet. One moment, Mrs. Zhigalov, please come here. [*Takes* MRS. ZHIGALOV *on one side, panting.*] Listen. The General's on his way, I've got hold of one at last. Had a terrible time. It's a real General, very dignified, very old. He must be about eighty—ninety, even.

MRS. ZHIGALOV. But when will he get here?

NYUNIN. Any moment now. You'll be grateful to me all your life. He's a General and a half, a regular conquering hero! Not some wretched footslogging old square-basher, this—he's a General of the Fleet. His rank's Commander, and in naval lingo that's equal to a Major-General, or an Under-Secretary in the civil service, there's nothing in it. It's higher, in fact.

MRS. ZHIGALOV. You're not pulling my leg are you, Andrew?

NYUNIN. What—think I'd swindle you, eh? Set your mind at rest.

MRS. ZHIGALOV [*sighing*]. I don't want us to waste our money, Andrew.

NYUNIN. Don't worry. He's a cracking good General! [*Raising his voice.*] I said to him: 'You've quite forgotten us, sir,' says I. 'You shouldn't forget old friends, sir. Mrs. Zhigalov's very annoyed with you,' I say. [*Goes to the table and sits down.*] 'But look here, young man,' says he, 'how can I go when I don't know the groom?' 'Oh really, sir, don't stand on ceremony.' 'The groom's a grand chap,' I tell him. 'Good mixer and all that. Works for a pawnbroker, writes out the price-tickets,' says I. 'But don't think he's some awful little tick, sir, or a frightful bounder. Respectable ladies serve in pawn-shops nowadays.' He slaps me on the shoulder, we each smoke a Havana, and now he's on his way. Wait, all of you, don't start eating.

APLOMBOV. But when will he get here?

NYUNIN. Any moment. He was putting on his galoshes when I left. Wait a bit, everyone, don't start.

APLOMBOV. Then you'd better tell them to play a march.

NYUNIN [*shouts*]. Hey, you in the band! A march!

[*The band plays a march for a minute.*]

WAITER [*announces*]. Mr. Revunov-Karaulov!

[ZHIGALOV, MRS. ZHIGALOV *and* NYUNIN *run to meet him. Enter* REVUNOV-KARAULOV.]

MRS. ZHIGALOV [*bowing*]. Make yourself at home, sir. Pleased to meet you, I'm sure.

REVUNOV. Delighted!

ZHIGALOV. We're plain, ordinary, humble folk, sir, but don't imagine we'd go in for any funny business. We set great store by nice people here, nothing's too good for them. So make yourself at home.

REVUNOV. Delighted indeed!

NYUNIN. May I introduce, General? This is the bridegroom, Mr. Epaminondas Aplombov and his newly-born—that is, his newly-wedded wife. Mr. Ivan Yat, who works in the telegraph office. Mr. Kharlampy Dymba, a foreigner of Greek extraction who's in the confectionery line. Mr. Osip Babelmandebsky. And so on and so forth. The rest are pretty small beer. Sit down, sir.

REVUNOV. Delighted! Excuse me, ladies and gentlemen, I must have a word with young Andrew. [*Takes* NYUNIN *aside.*] I feel a bit awkward, my boy. Why all this 'General' stuff? It's not as if I was one—I'm a naval Commander, and that's even lower than a Colonel.

NYUNIN [*speaks into his ear as if he was deaf*]. I know, but please let us call you General, Commander. This is an old-fashioned family, see. They look up to their betters, they like to show due respect.

REVUNOV. Oh, in that case of course. [*Going to the table.*] Delighted!

MRS. ZHIGALOV. Do sit down, General. Be so kind! And have something to eat and drink, sir. But you must excuse us, you being used to fancy things, like—we're plain folk, we are.

REVUNOV [*not hearing*]. What's that? I see. Very well. [*Pause.*] Very well. In the old days people all lived the simple life and were content. I live simply too, for all my officer's rank. Young Andrew comes to see me today and asks me to this wedding. 'How can I go,' I ask, 'when I don't know 'em? It's rather awkward.' 'Well,' says he, 'they're old-fashioned folk with no frills, always glad to have someone drop in.' Well, of course, if that's the way of it—why not? Delighted! It's boring being on your own at home, and if having me at a wedding can please anyone, they're only too welcome, say I.

ZHIGALOV. So it was out of the kindness of your heart, General? I hand it to you. I'm a plain man too, I don't hold with any sort of funny business, and I respect people like myself. Help yourself, sir.

APLOMBOV. Have you been retired long, sir?

REVUNOV. Eh? Yes, yes, I have. Quite true. Yes. I say, look here, what's all this? The herring tastes sour, and so does the blasted bread.

ALL. Blasted bread? Bless the bride! To the bride and groom!

[APLOMBOV *and* DASHA *kiss.*]

REVUNOV. Tee hee hee! Good health! [*Pause.*] Yes, in the old days things were straightforward and everyone was content. I like things shipshape. I'm an old man, after all—retired back in 'sixty-five! I'm seventy-two. Yes. There were times in the old days when they liked to cut a bit of a dash, of course. [*Seeing* MOZGOVOY.] You, er—a sailor, are you?

MOZGOVOY. Yessir.

REVUNOV. Aha! I see. Yes, the navy's always been a pretty rugged service—makes a man think and cudgel the old brain. Every little word has its special kind of meaning. For instance: 'Topmen aloft! To the foresail and mainsail yards!' What does that mean? You can bet your sailor knows! Tee hee hee! It's as tricky as geometry!

NYUNIN. To the health of General Theodore Revunov-Karaulov!

[*The band plays a flourish. Cheers.*]

YAT. Now, sir, you've just been telling us about the difficulties of naval service. But telegraphing's no easier, you know. Nowadays, sir, no one can get a job on the telegraph unless he can read and write French and German. But the hardest thing we do is sending morse. It's a very tough job! Listen to me, sir. [*Bangs his fork on the table in imitation of someone sending morse.*]

REVUNOV. What does that mean?

YAT. It means: 'I much respect you, sir, for your distinguished qualities.' Think it's easy, eh? Here's some more. [*Taps.*]

REVUNOV. Make it a bit louder, I can't hear.

YAT. That means: 'Madam, how happy I am to hold you in my arms.'

REVUNOV. Who's this 'madam'? Yes—. [*To* MOZGOVOY.] Now, if you're running before a good breeze and want to, er, set your topgallants and royals, you must order: 'Hands aloft to the topgallants and royals!' And while they cast loose the sails on the yards, down on deck they're manning the topgallant and royal sheets, halyards and braces.

BEST MAN [*getting up*]. Ladies and gent——

REVUNOV [*interrupting*]. Yes indeed. We've plenty of words of command, we have that. 'In on the topgallant and royal sheets! Haul taut the halyards!' Not bad, eh? But what does it all mean, what's the sense of it? It's all very simple. They haul the topgallant and royal sheets, see? And lift the halyards all at once, squaring off the royal sheets and royal halyards as they hoist, at the same time easing the braces from the sails by the required amount. So when the sheets are taut and the halyards all run right up, the topgallants and royals are drawing and the yards are braced according to the wind direction——

NYUNIN [*to* REVUNOV]. Our hostess asks, would you mind talking about something else, sir? The guests can't make this out, and they're bored.

REVUNOV. What! Who's bored? [*To* MOZGOVOY.] Young man! Now, if she's close-hauled under full sail on the starboard tack and you have to wear ship—what command must you give? Why, pipe all hands on deck! Wear ship! Tee hee hee!

NYUNIN. That's quite enough, Commander. Have something to eat.

REVUNOV. As soon as they all run up, the command's given at once: 'Stand by to wear ship!' What a jolly life! As you give the commands you watch the sailors run to their stations like greased lightning and unfurl the topgallants. You can't help shouting: 'Well done, lads!' [*Chokes and coughs.*]

BEST MAN [*hastens to take advantage of the ensuing pause*]. This evening, as we are, so to speak, gathered together to honour our dear——

REVUNOV [*interrupting*]. Yes, there's all that to remember, you know! For instance—let fly the foresheet, let fly the main!

BEST MAN [*offended*]. Why does he keep interrupting? We shan't get through a single speech at this rate.

MRS. ZHIGALOV. We're poor benighted folk, sir, we can't make sense of all that. Why not tell us something about——

REVUNOV [*not hearing*]. I've already eaten, thanks. Goose, you say? No, thank you. Yes, it all comes back to me—what jolly times those were, boy! You sail the seas without a care in the world, and [*in a trembling voice*]—remember the excitement of tacking? What sailor isn't fired by the thought of that manœuvre? Why, as soon as the command rings out, 'Pipe all hands on deck! Ready about!'—it's as if an electric spark ran through them all. Everyone, from the captain to the last sailor, is galvanized!

MRS. ZMEYUKIN. Oh, what a bore!

[*A general murmur of complaint.*]

REVUNOV [*not hearing*]. Thanks, I've already eaten. [*Carried away.*] Everyone stands by, and all eyes are glued on Number One. 'Haul taut the fore and main starboard braces and the port mizzen top braces and counterbraces on the port side!' shouts Number One. It's all done in a flash. 'Let fly the fore-sheet, let fly the jib-sheet! Hard a' starboard!' [*Stands up.*] She comes up into the wind and at last the sails start flapping. 'The braces! Look alive on those braces!' yells Number One. His eyes are glued to the main topsail, then at last that starts flapping too—so she's started to come about, and the command is given like a crack of thunder: 'Let go the main top bowline! Pay out the braces!' Then everything flies and cracks, a regular pandemonium, and it's all been done to perfection. We've brought her about!

MRS. ZHIGALOV [*flaring up*]. And you a General! A hooligan, more like! You should be ashamed, at your time of life!

REVUNOV [*mishearing*]. A slice of tripe? Thanks, I don't mind if I do.

MRS. ZHIGALOV [*loudly*]. I say you should be ashamed at your age! Call yourself a General, behaving like that!

NYUNIN [*embarrassed*]. Oh, look, everyone—why all the fuss? Really——

REVUNOV. Firstly, I'm not a General—I'm a naval Commander, which is equal to Lieutenant-Colonel in the army.

MRS. ZHIGALOV. If you aren't a General, why take the money? We didn't pay good money for you to break up the happy home!

REVUNOV [*bewildered*]. Pay what money?

MRS. ZHIGALOV. You know very well what money. You got your twenty-five roubles from Mr. Nyunin right enough. [*To* NYUNIN.] You should be ashamed of yourself, Andrew. I didn't ask you to hire something like this!

NYUNIN. Oh, I say, cut it out. Why make a fuss?

REVUNOV. Hired? Paid? What do you mean?

APLOMBOV. I say, just one moment—you did receive twenty-five roubles from Andrew Nyunin, I take it?

REVUNOV. Twenty-five roubles? [*The truth dawns.*] So that's it! I see! What a rotten, dirty trick!

APLOMBOV. Well, you took the money, I gather?

REVUNOV. I took no money. Get away from me! [*Stands up from the table.*] What a filthy, rotten trick! To insult an old sailor, an officer who's seen honourable service. If this was a respectable house I could challenge someone to a duel, but what can I do here? [*In despair.*] Where's the door? Which is the way out? Waiter, take me out! Waiter! [*Moves off.*] What a filthy, rotten trick! [*Goes out.*]

MRS. ZHIGALOV. But where are the twenty-five roubles, Andrew?

NYUNIN. Oh, what a lot of fuss about nothing! A fat lot that matters, with everyone enjoying themselves, damn it—I don't know what you're talking about. [*Shouts.*] To the health of bride and groom! You in the band! Play a march! Band! [*The band plays a march.*] To the bride and groom!

MRS. ZMEYUKIN. I'm choking. Let me have air! I choke when I'm near you.

YAT [*ecstatically*]. Superb creature!

[*A lot of noise.*]

BEST MAN [*trying to shout them down*]. Ladies and gentlemen! On this, er, so to speak, day——

CURTAIN

THE ANNIVERSARY

[*Юбилей*]

A FARCE IN ONE ACT

(1891)

CHARACTERS

ANDREW SHIPUCHIN, chairman of the board of a mutual credit bank, a man in his forties who wears a monocle

TATYANA, his wife, aged 25

KUZMA KHIRIN, an elderly bank clerk

MRS. NASTASYA MERCHUTKIN, an old woman who wears an old-fashioned overcoat

Shareholders and bank clerks

The action takes place at the offices of the N. Mutual Credit Bank

The Chairman's office. A door, left, leading into the main office. Two desks. The furnishings have pretensions to extreme luxury: velvet-upholstered furniture, flowers, statues, carpets and a telephone.

Noon.

KHIRIN *is alone. He wears felt boots.*

KHIRIN [*shouts through the door*]. Send someone to get fifteen copecks' worth of valerian drops from the chemist's and have some fresh water brought into the Chairman's office. Must I tell you a hundred times? [*Goes towards the desk.*] I'm all in. I've been writing for over seventy-two hours on end without a wink of sleep. I write here all day, I write at home all night. [*Coughs.*] I ache all over, what's more. I've a feverish chill and a cough, my legs ache and I keep seeing stars or exclamation marks or something. [*Sits down.*] And today that swine, that simpering buffoon—our Chairman, I mean—is to speak at the General Meeting on 'Our Bank: its Present and its Future'. He doesn't half fancy himself, I must say! [*Writes.*] Two . . . one . . . one . . . six . . . oh . . . seven. Then—six . . . oh . . . one . . . six. He only wants to put over a lot of eyewash, and I'm to sit here working my fingers to the bone, if you please! Fills his speech with highfalutin' poppycock and expects me to tot up figures day in day out, God damn and blast him! [*Clicks his counting frame.*] I can't stand it. [*Writes.*] Ah well—one . . . three . . . seven . . . two . . . one . . . oh. He's promised to see I don't lose by it. If all goes well this afternoon and he manages to bamboozle his audience, he's promised me a gold medal and a three-hundred-rouble bonus. We shall see. [*Writes.*] But if I get nothing for my pains, my lad, then you can watch out—I'm apt to fly off the handle! You put my back up, chum, and you'll find yourself in Queer Street, believe you me!

[*Noise and applause off-stage.* SHIPUCHIN*'s voice:* 'Thank you, thank you. Most touched.' SHIPUCHIN *comes in. He wears evening dress with a white tie and carries an album which has just been presented to him.*]

SHIPUCHIN [*standing in the doorway and turning to the main office*]. Dear colleagues, I shall treasure this gift of yours till my dying day as a memento of the happiest years of my life. Yes, gentlemen. Thank you again. [*Blows a kiss and goes towards* KHIRIN.] My dear, good Khirin.

[*While he is on stage, clerks come and go from time to time with papers for him to sign.*]

KHIRIN [*getting up*]. Mr. Shipuchin, may I felicitate you on the Bank's fifteenth anniversary and wish——

SHIPUCHIN [*shakes his hand vigorously*]. Thank you very much, my dear fellow. Thank you. On this auspicious occasion we may, I think, embrace. [*They kiss.*] So glad, so glad. Thank you for your services, thank you for everything, everything. If I've ever done anything useful since I had the honour to be Chairman of the Bank, I owe it mainly to my colleagues. [*Sighs.*] Yes, my dear fellow, fifteen years! Fifteen years, or my name's not Shipuchin! [*Briskly.*] Well, how's my speech? Is it coming along?

KHIRIN. Yes, I've only about five pages to do.

SHIPUCHIN. Fine. So it will be ready by three o'clock, will it?

KHIRIN. It will if I'm not interrupted, there's only a bit to do.

SHIPUCHIN. Marvellous. Marvellous, or my name's not Shipuchin. The General Meeting's at four. Tell you what, old man—let me have the first part to look through. Come on, hurry up. [*Takes the speech.*] I'm hoping for a lot from this speech. It's a statement of faith—a firework display, rather. There's some pretty hot stuff in this, or my name's not Shipuchin. [*Sits down and reads the speech to himself.*] I'm dead beat, my word I am. I had a spot of gout last night, I've been rushed off my feet doing errands all morning, and now there's all this excitement, applause and fuss. I'm worn out.

KHIRIN [*writes*]. Two . . . oh . . . oh . . . three . . . nine . . . two . . . oh. These figures make me dizzy. Three . . . one . . . six . . . four . . . one . . . five. [*Clicks his counting-frame.*]

SHIPUCHIN. There was some more unpleasantness too. Your wife was here this morning, complaining about you again—said you ran after her with a knife last night, and your sister-in-law too. Whatever next, Khirin! This won't do.

KHIRIN [*sternly*]. Mr. Shipuchin, may I venture to ask you a favour on this anniversary occasion, if only out of consideration for the drudgery I do here? Be so good as to leave my family life alone, would you mind?

SHIPUCHIN [*sighs*]. You're quite impossible, Khirin. You're a very decent, respectable fellow, but with women you're a regular Jack the Ripper. You are, you know. I can't see why you hate them so.

KHIRIN. Well, I can't see why you like them so. [*Pause.*]

SHIPUCHIN. The clerks have just given me an album, and I hear the shareholders want to present me with an address and a silver tankard. [*Playing with his monocle.*] Very nice, or my name's not Shipuchin—no harm in it at all. A little ceremony's needed for the sake of the Bank's reputation, damn it. You're one of us, so you're in the know, of course. I wrote the address myself—and as for the tankard, well, I bought that too. Yes, and it set me back forty-five roubles to have the address bound, but there was nothing else for it. *They'd* never have thought of it. [*Looks round.*] What furniture and fittings! Not bad, eh? They call me fussy—say I only care about having my door-handles polished, my clerks turned out in smart neck-ties and a fat commissionaire standing at my front door. Not a bit of it, sirs. Those door-handles and that commissionaire aren't trifles. At home I can behave like some little suburban tyke—sleep and eat like a hog, drink like a fish——

KHIRIN. I'll thank you not to make these insinuations.

SHIPUCHIN. Oh, no one's insinuating anything, you really are impossible! Now, as I was saying, at home I can be a jumped-up little squirt with my own nasty little habits. But *here* everything must be on the grand scale. This is a bank, sir! Here every detail must impress and wear an air of solemnity, as you might say. [*Picks up a piece of paper from the floor and throws it on the fire.*] My great merit is simply that I've raised the Bank's prestige. Tone's a great thing. A great thing is tone, or my name's not Shipuchin. [*Looking* KHIRIN *over.*] My dear fellow, a shareholders' deputation may come in any moment, and here are you in those felt boots and that scarf and, er, that jacket thing—what a ghastly colour! You might have worn tails or at least a black frock-coat——

KHIRIN. My health matters more to me than your shareholders. I feel sore all over.

SHIPUCHIN [*excitedly*]. Well, you must admit you look an awful mess. You're spoiling the whole effect.

KHIRIN. If the deputation comes, I can make myself scarce. What a fuss about nothing. [*Writes.*] Seven . . . one . . . seven . . . two . . .

one . . . five . . . oh. I can't stand messes either. Seven . . . two . . . nine. [*Clicks his counting-frame.*] I don't like messes and muddles, which is why I wish you hadn't invited women to tonight's celebration dinner.

SHIPUCHIN. Oh, don't talk rot.

KHIRIN. I know you'll fill the place with women tonight just for the look of the thing, but you watch out—they'll muck everything up for you. Mischief and trouble—it all comes from women.

SHIPUCHIN. Not at all. Ladies' company is elevating.

KHIRIN. It is, is it? Your wife's an educated woman, I think, but last Monday she blurted out something—well, it took me two days to get over it. She suddenly asks in front of some strangers: 'Is it true my husband bought some shares in the Dryazhsky-Pryazhsky business for our bank, and then the price came down on the stock exchange? Oh, my husband's so worried!' she says. Before total outsiders! Why do you confide in them? That's what I don't see. Want them to land you in clink?

SHIPUCHIN. Oh, do cut it out, this is all much too gloomy for a celebration. By the way, you just reminded me. [*Looks at his watch.*] My dear wife's due here any moment. I really should have popped over to the station to meet the poor thing, but I'm too busy and—and too tired. And the fact is, I'm not all that keen on her coming. Actually, I am quite keen, but I'd rather she stayed on at her mother's for a day or two. She'll insist I spend the whole evening with her, when we were planning a little trip after dinner. [*Gives a start.*] I say, I'm getting as nervous as a kitten—I'm so on edge, I feel I'll burst into tears at the slightest provocation. Yes, one must be firm, or my name's not Shipuchin.

[TATYANA *comes in, wearing a mackintosh, with a travelling handbag slung over her shoulder.*]

SHIPUCHIN. I say! Talk of the devil!

TATYANA. Darling! [*Runs to her husband; a prolonged kiss.*]

SHIPUCHIN. Why, we were just talking about you. [*Looks at his watch.*]

TATYANA [*out of breath*]. Have you missed me? How are you? I haven't been home yet, I came straight here from the station. I've such a lot to tell you, I can hardly wait. I won't take my coat off, I'm just going. [*To* KHIRIN.] Good afternoon, Mr. Khirin. [*To her husband.*] Is all well at home?

SHIPUCHIN. Yes. And you've grown plumper and prettier during the week. Now, how was the trip?

TATYANA. Fine. Mother and Katya send their love. Vasily told me to give you a kiss. [*Kisses him.*] Aunty's sent a pot of jam and everyone's angry with you for not writing. Zina sends a kiss too. [*Kisses him.*] Oh, if you only knew what's been going on, if you had any idea! I'm scared to talk about it, really. What a business! But you don't look very pleased to see me.

SHIPUCHIN. On the contrary—. Darling——

[*Kisses her.* KHIRIN *gives an angry cough.*]

TATYANA [*sighs*]. My poor, poor Katya! I'm so, so sorry for her!

SHIPUCHIN. My dear, we're celebrating our anniversary today, and a shareholders' deputation may turn up any moment. And you're not properly dressed.

TATYANA. But of course! The anniversary! Congratulations, gentlemen. I wish you—. So it's the day of the meeting and the dinner. I adore all that. Remember that lovely shareholders' address? It took you such a long time to write. Are they going to present it today?

[KHIRIN *coughs angrily.*]

SHIPUCHIN [*embarrassed*]. Darling, one doesn't talk about such things. Now, look—why don't you go home?

TATYANA. Yes, yes, of course. It won't take a moment to tell you, and then I'll go. I'll tell you it all from the beginning. Now then, after you'd seen me off, I sat next to that stout lady, remember? And I started reading. I don't like talking in trains, so I went on reading for three stops, and not a word did I utter to a soul. Then evening came on and I began to feel depressed, see? There was this dark-haired young fellow sitting opposite me—not bad-looking, quite attractive, actually. Well, we got talking. A sailor came along and some student or other. [*Laughs.*] I told them I wasn't married. Oh, they were all over me! We chattered away till midnight—the dark young man told some screamingly funny stories and the sailor kept singing! I laughed till my sides ached. And when the sailor—

oh, those sailors!—when it came out that I was called Tatyana, do you know what he sang? [*Sings in a bass voice.*]

'Onegin, how can I deny
I'll love Tatyana till I die?'

[*Roars with laughter.* KHIRIN *coughs angrily.*]

SHIPUCHIN. I say, Tatyana, we're annoying Mr. Khirin. You go home, darling. Tell me about it later.

TATYANA. Never mind, never mind, let him listen too, he'll find it very interesting. I'll be through in a moment. Sergey met me at the station. And then another young man turned up, a tax inspector, I think—quite nice-looking, charming in fact, especially his eyes. Sergey introduced us and the three of us went off together. The weather was glorious.

[*Voices off-stage*: 'You mustn't go in there. What do you want?' MRS. MERCHUTKIN *comes in.*]

MRS. MERCHUTKIN [*in the doorway, waving people off*]. You take your hands off of me, I never heard of such a thing! I want to see His Nibs. [*Coming in, to* SHIPUCHIN.] May I introduce myself, sir? Nastasya Merchutkin's the name, and my husband's in the civil service, like.

SHIPUCHIN. And what can I do for you?

MRS. MERCHUTKIN. Well, it's this way, sir. My husband, him that's in the service, has been real poorly these five months, and while he was laid up at home under the doctor he was given the sack for no reason, sir, and when I went to get his pay, do you know what they'd done? Gone and docked him twenty-four roubles thirty-six copecks, they had. What for? That's what I'd like to know. 'He borrowed it from the kitty,' they tell me, 'and other people guaranteed the loan.' What an idea—as if he'd borrow money without asking me first! They can't do this to me, sir. I'm a poor woman, I only manage by taking lodgers. I'm a weak, defenceless woman. Everyone insults me, no one says a kind word to me.

SHIPUCHIN. Very well then. [*Takes her application and reads it standing up.*]

TATYANA [*to* KHIRIN]. Well, I must begin at the beginning. Last week I suddenly get a letter from Mother. She writes that my sister Katya's had a proposal from a certain Grendilevsky—a very nice, modest young man, but with no means or position at all. Now, by rotten bad luck Katya was rather gone on him, believe it or not. So what's to be done? Mother writes and tells me to come at once and influence Katya.

KHIRIN [*sternly*]. Look here, you're putting me off. While you go on about Mother and Katya, I've lost my place and I'm all mixed up.

TATYANA. Well, it's not the end of the world! And you listen when a lady talks to you! Why so peeved today? Are you in love? [*Laughs.*]

SHIPUCHIN [*to* MRS. MERCHUTKIN]. I say, look here, what's all this about? I can't make sense of it.

TATYANA. In love, eh? Aha—blushing, are we?

SHIPUCHIN [*to his wife*]. Tatyana, go into the office for a moment, dear. I won't be long.

TATYANA. All right. [*Goes out.*]

SHIPUCHIN. It makes no sense to me. You've obviously come to the wrong address, madam—your business is really no concern of ours. You should apply to the institution where your husband was employed.

MRS. MERCHUTKIN. I've been in half a dozen different places already, mister, and they wouldn't even listen. I was at my wits' end when Boris, that's my son-in-law, has the bright idea of sending me to you. 'You go to Mr. Shipuchin, Mum,' says he. 'The gentleman carries a lot of weight, and he can do anything.' Please help me, sir!

SHIPUCHIN. Mrs. Merchutkin, there's nothing we can do for you. Look here—as far as I can see, your husband worked for the War Office medical department, but this establishment is a completely private business. This is a bank, can't you see?

MRS. MERCHUTKIN. Now, about my husband's illness, sir, I have a doctor's certificate. Here it is, if you'll kindly have a look——

SHIPUCHIN [*irritatedly*]. All right, I believe you. But I repeat, it's no concern of ours.

[TATYANA'*s laughter is heard off-stage, followed by a man's laughter.*]

SHIPUCHIN [*glancing at the door*]. She's stopping them working out there. [*To* MRS. MERCHUTKIN.] This is odd—a bit funny, really. Surely your husband knows where to apply?

MRS. MERCHUTKIN. The poor man knows nothing, sir. 'It's none of your business,' he keeps saying. 'So hop it!' I can't get another word out of him.

SHIPUCHIN. I repeat, madam. Your husband was employed by the War Office medical department, but this is a bank, a private business.

MRS. MERCHUTKIN. Quite so, quite so. I understand, mister. Then tell them to give me fifteen roubles, say. I don't mind waiting for the rest, sir.

SHIPUCHIN [*sighs*]. Phew!

KHIRIN. Mr. Shipuchin, I'll never finish the speech at this rate.

SHIPUCHIN. Just a moment. [*To* MRS. MERCHUTKIN.] You won't listen to reason. But you must see—it's as absurd for you to come to us with an application like this as it would be for someone to try and get a divorce at—let's say at a chemist's shop or the Assay Office.

[*A knock on the door.*]

TATYANA [*off-stage*]. Can I come in, Andrew?

SHIPUCHIN [*shouts*]. Just a moment, darling. [*To* MRS. MERCHUTKIN.] They underpaid you, but what business is that of ours? Besides, this is our anniversary celebration, madam, we're busy—and someone may come in any moment. Please excuse me.

MRS. MERCHUTKIN. Pity a helpless orphan, sir. I'm a weak, defenceless woman. Fair worried to death I am, what with lodgers to have the law on, my husband's affairs to handle, a house to run—and my son-in-law out of work as well.

SHIPUCHIN. Mrs. Merchutkin, I—. No, I'm sorry, I can't talk to you. My head's going round and round. You're annoying us, besides wasting your own time. [*Sighs, aside.*] The woman's daft, or my name's not Shipuchin. [*To* KHIRIN.] Mr. Khirin, would you mind explaining to Mrs. Merchutkin? [*Makes a gesture of despair and goes off into the board-room.*]

KHIRIN [*goes up to* MRS. MERCHUTKIN, *sternly*]. What do you want?

MRS. MERCHUTKIN. I'm a weak, defenceless woman. I may look strong, but if you took me to pieces you wouldn't find one healthy vein in my body. I can hardly stand, and my appetite's gone. I drank some coffee this morning, but it didn't go down well at all, it didn't.

KHIRIN. I asked you a question: what is it you want?

MRS. MERCHUTKIN. Just tell them to pay me fifteen roubles, mister, and the rest in a month, say.

KHIRIN. But I thought you'd already been told in words of one syllable. This is a bank!

MRS. MERCHUTKIN. Yes, yes, of course. And I can show you the doctor's certificate if you want.

KHIRIN. Have you taken leave of your senses, or what?

MRS. MERCHUTKIN. I'm only asking for my rights, kind sir. I don't want what isn't mine.

KHIRIN. I'm asking you if you've taken leave of your senses, madam. Oh, damn and blast me, I've no time to bandy words with you, I'm busy. [*Points to the door.*] Would you mind?

MRS. MERCHUTKIN [*astonished*]. But what about my money?

KHIRIN. The fact is, you've no senses to take leave of. This is your trouble. [*Taps his finger on the table and then on his forehead.*]

MRS. MERCHUTKIN [*taking offence*]. What! Who do you think you are! You do that to your own wife. My husband's in the executive branch, so don't you try your tricks on me.

KHIRIN [*losing his temper, in a low voice*]. Get out of here!

MRS. MERCHUTKIN. Cor, hark at him! You mind what you're saying!

KHIRIN [*in a low voice*]. If you don't leave this instant I'll send for the porter. Get out. [*Stamps.*]

MRS. MERCHUTKIN. Who do you think you're talking to? You don't scare me, I know your sort! You're crackers.

KHIRIN. I don't think I've ever seen anything nastier in my life. Ugh, what a pain in the neck! [*Breathes heavily.*] I repeat, do you hear? If you won't clear out, I'll pulverize you, you old horror. I'm quite capable of crippling you for life, that's the sort of man I am. I'll stop at nothing.

MRS. MERCHUTKIN. You're all bark and no bite. You don't scare me, I know your sort.

KHIRIN [*in despair*]. I can't stand the sight of her. Sickening! This is too much! [*Goes to the table and sits down.*] They unleash a horde of women in the bank and I can't get on with the speech. It's too much.

MRS. MERCHUTKIN. I'm not asking for what isn't mine. I want my rights, I do. Cor, look at him! Sits around in the office with his felt boots on! Cheek! Where was you brought up?

[SHIPUCHIN *and* TATYANA *come in.*]

TATYANA [*coming in after her husband*]. We went to a party at the Berezhnitskys'. Katya was wearing a dear little blue silk frock with an open neck, and trimmed with fine lace. She does look nice with her hair up, and I arranged it myself. Her dress and hair were quite devastating!

SHIPUCHIN [*who now has migraine*]. Yes, yes, I'm sure. Someone may come in here any moment.

MRS. MERCHUTKIN. Sir!

SHIPUCHIN [*despondently*]. What now? What do you want?

MRS. MERCHUTKIN. Sir! [*Points to* KHIRIN.] This man here, this creature—he taps his forehead at me and then on the table. You tell him to look into my case, but he sneers at me and makes nasty remarks. I'm a weak, defenceless woman, I am.

SHIPUCHIN. Very well, madam, I'll see about it. I'll take steps. Now do go, I'll deal with it later. [*Aside.*] My gout's coming on.

KHIRIN [*goes up to* SHIPUCHIN, *quietly*]. Mr. Shipuchin, let me send for the hall-porter and have her slung out on her ear. This beats everything.

SHIPUCHIN [*terrified*]. No, no! She'll only raise Cain and there are a lot of private apartments in this block.

MRS. MERCHUTKIN. Sir!

KHIRIN [*in a tearful voice*]. But I have a speech to write. I shan't get it done in time. [*Goes back to the desk.*] I can't stand this.

MRS. MERCHUTKIN. Please sir, when do I get my money? I need it at once.

SHIPUCHIN [*aside, indignantly*]. What a perfectly horrible old bitch! [*To her, gently.*] As I've said already, madam, this is a bank—a private business establishment.

MRS. MERCHUTKIN. Have mercy, kind sir. Think of yourself as my father. If the doctor's certificate isn't enough, I can bring a paper from the police too. Tell them to pay me the money.

SHIPUCHIN [*sighs heavily*]. Phew!

TATYANA [*to* MRS. MERCHUTKIN]. I say, old girl, you're in the way, do you hear? This won't do, you know.

MRS. MERCHUTKIN. Pretty lady, I've no one to stick up for me. Food and drink don't mean a thing, dearie, and I had some coffee this morning, but it didn't go down well at all, it didn't.

SHIPUCHIN [*exhausted, to* MRS. MERCHUTKIN]. How much do you want?

MRS. MERCHUTKIN. Twenty-four roubles, thirty-six copecks.

SHIPUCHIN. Very well then. [*Gets twenty-five roubles out of his wallet and gives it to her.*] Here's your twenty-five roubles. Take it and be off with you.

[KHIRIN *coughs angrily.*]

MRS. MERCHUTKIN. Thanking you kindly, sir. [*Puts the money away.*]

TATYANA [*sitting down near her husband*]. I must be off home, though. [*Glances at her watch.*] But I haven't finished yet. I'll only be a minute, and then I'll go. What a business, though, what a to-do! Well, off we go to the Berezhnitskys' party. It was all right, quite amusing, if nothing special. Katya's young man Grendilevsky was there too, of course. Well, I have a word with Katya, I shed a few tears, and I use my influence on her, so she has it out with Grendilevsky then and there at the party and turns him down. Well, thinks I, this has all gone like clockwork—I've set Mother's mind at rest, I've saved Katya, and now I can relax. But what do you think? Katya and I are walking in the garden just before supper, when suddenly—. [*Excitedly.*] When suddenly we hear a shot! No, I can't talk about it calmly. [*Fans herself with her handkerchief.*] It's too much for me!

[SHIPUCHIN *sighs.*]

TATYANA [*weeps*]. We rush to the summer-house, and there—there lies poor Grendilevsky with a pistol in his hand.

SHIPUCHIN. Oh, I can't stand this—can't stand it, I tell you! [*To* MRS. MERCHUTKIN.] What more do you want?

MRS. MERCHUTKIN. Please sir, can my husband have his job back?

TATYANA [*weeping*]. He shot himself straight through the heart, just here. Katya fainted, poor dear. And he got the fright of his life. He just lies there and asks us to send for the doctor. The doctor turns up quite soon and—and saves the poor boy.

MRS. MERCHUTKIN. Please sir, can my husband have his job back?

SHIPUCHIN. Oh, this is *too* much! [*Weeps.*] I can't cope! [*Stretches out both hands to* KHIRIN, *in despair.*] Get rid of her, chuck her out, for God's sake!

KHIRIN [*going up to* TATYANA]. You clear out!

SHIPUCHIN. No, not her, this one—this monster. [*Points to* MRS. MERCHUTKIN.] This one!

KHIRIN [*not understanding him, to* TATYANA]. Buzz off! [*Stamps.*] Beat it!

TATYANA. What! *What* did you say? Are you stark, staring mad?

SHIPUCHIN. This is ghastly! Oh, I'm so unhappy! Get her out of here, I tell you.

KHIRIN [*to* TATYANA]. You get out of here, or I'll cripple you for life! Mutilate you, I will. I'll stick at nothing!

TATYANA [*runs away from him, as he chases her*]. How dare you! I like your cheek! [*Shouts.*] Andrew, save me! Andrew! [*Shrieks.*]

SHIPUCHIN [*runs after them*]. Stop that, please. Be quiet. Have a heart!

KHIRIN [*chases* MRS. MERCHUTKIN]. Get out of here! Get hold of her, beat her, slit her throat!

SHIPUCHIN [*shouts*]. Stop that, please, I beg you!

MRS. MERCHUTKIN. Oh, goodness me! [*Shrieks.*] Heavens!

TATYANA [*shouts*]. Save me, save me! Oh, oh, I feel faint! Faint, I tell you! [*Jumps on a chair, then falls on the sofa and groans as if in a faint.*]

KHIRIN [*chases* MRS. MERCHUTKIN]. Get hold of her, thrash her, slit her throat!

MRS. MERCHUTKIN. Oh dearie me, I've come over all queer. Oh! [*Faints in* SHIPUCHIN*'s arms. There is a knock on the door, and a voice off-stage*: 'The deputation!']

SHIPUCHIN. Deputation, reputation, occupation——

KHIRIN [*stamps*]. Get out of here, God damn me! [*Rolls up his sleeves.*] Let me get at her! I may commit an atrocity!

[*Enter a deputation of five men, all in evening dress. One carries the address bound in velvet, another the tankard. Bank employees look through the office door.* TATYANA *is on the sofa,* MRS. MERCHUTKIN *is in* SHIPUCHIN*'s arms, and both are quietly groaning.*]

A SHAREHOLDER [*reads in a loud voice*]. Mr. Shipuchin, our dear and most respected friend. As we turn a retrospective glance on the history of this financial establishment, running the mind's eye over the story of its gradual evolution, the impression we receive is highly gratifying. In the early stages of the Bank's existence the limited extent of our basic capital and the absence of any serious operations, as also the vagueness of our aims, did, it is true, confront us with Hamlet's question: 'To be or not to be?' At one time voices were even heard in favour of closing the Bank. But then *you* became head of the establishment. Your knowledge, energy and natural tact have accounted for our tremendous success and exceptional prosperity. The reputation of the Bank [*coughs*], the Bank's reputation——

MRS. MERCHUTKIN [*groans*]. Oh, oh!

TATYANA [*groans*]. Water, water!

SHAREHOLDER [*continues*]. The reputation—. [*Coughs.*] The Bank's reputation has been raised by you to such heights that our establishment can now compete with the best foreign establishments——

SHIPUCHIN. Deputation, reputation, occupation——.

'Two friends one night went for a walk,
And on that walk they had a talk.'

'Oh, tell me not your young life's ruined
And poisoned by my jealousy.'

SHAREHOLDER [*continues in embarrassment*]. And now, dear respected friend, casting the eye of objectivity at the present, we, er—. [*Lowering his voice.*] Under the circumstances we'd better come back later. Later.

[*They go out in embarrassment.*]

CURTAIN

SMOKING IS BAD FOR YOU

[*О вреде табака*]

A MONOLOGUE IN ONE ACT

(1903)

CHARACTER

IVAN NYUKHIN, a hen-pecked husband whose wife keeps a music school and a girls' boarding-school

The stage represents the platform in the hall of a provincial club

NYUKHIN *struts in majestically. He has long side-whiskers, his upper lip is clean-shaven, and he wears an old, worn, tail-coat. He bows and adjusts his waistcoat.*

NYUKHIN. Ladies and er, in a manner of speaking, gentlemen. [*Strokes his side-whiskers.*]

It's been suggested to the wife that I should lecture here in aid of charity on some topic of general interest. I don't see why not. If I'm to lecture, I'll lecture—I just couldn't care less.

I'm not a professor, of course, and university degrees have passed me by. Still, for the last thirty years I've been working—non-stop, you might even say, ruining my health and all that—on problems of a strictly academic nature. I've done a lot of thinking, and even written some learned scientific articles, believe it or not—well, not exactly learned, but in the scientific line, as you might say. By the way, I wrote a great screed the other day on 'The Ill Effects Caused by Certain Insects'. My daughters really took to it, especially the bit about bed-bugs, but I read it through and tore it up. After all said and done and whatever you write, it comes down to good old Keating's Powder in the end, doesn't it? We've even got bugs in our piano. As the subject for my lecture today, I've chosen, as it were, the harmful effects of smoking on the human race. I'm a smoker myself, actually. But the wife told me to give today's lecture on why tobacco's bad for you, so what's the use of arguing? About tobacco as such, I just couldn't care less. But I suggest, ladies and gentlemen, that you attend to my present lecture with all due seriousness, or something worse may happen. If anyone's scared or put off by the idea of a dry, scientific lecture, he can stop listening and go. [*Adjusts his waistcoat.*]

I should like to ask the doctors in my audience to pay particular attention. My lecture is a mine of useful information for them, since nicotine not only has harmful effects, but is also used in medicine. For instance, if you put a fly in a snuff-box it will die—from a nervous breakdown, probably. Tobacco is, essentially, a plant. When I lecture my right eye usually twitches, but you can ignore that—it's pure nervousness. I'm a nervous wreck by and large, and this eye-twitching business started back in September 1889, on the thirteenth of the month—the very day when my wife gave birth, in a manner of speaking, to our fourth daughter Barbara. My daughters were all born on the 13th. Actually [*with a look at his watch*], time being short, let's not wander from the subject in hand.

My wife runs a school of music, I might add, and a private boarding-school—well, not a boarding-school exactly, but something in that line. Between you and me, my wife likes to complain of being hard up, but she's got a tidy bit salted away—a cool forty or fifty thousand—while I haven't a penny to my name, not a bean. But what's the use of talking? I'm the school matron. I buy food, keep an eye on the servants, do the accounts, make up exercise-books, exterminate bed-bugs, take the wife's dog for walks and catch mice. Last night I had the job of issuing flour and butter to Cook because we were going to have pancakes today. Well, this morning—to cut a long story short—when the pancakes are already cooked, in comes the wife to the kitchen to say three girls won't get any because they have swollen glands. It thus transpires that we have a pancake or two in hand. What are we to do with them? First my wife wants 'em put in the larder, then she changes her mind. 'You eat 'em, imbecile,' says she. That's what she calls me when she's in a bad mood—you imbecile, you snake, you hell-hound. Now how could anyone take me for a hell-hound? She's always in a bad mood, actually. Well, I didn't eat the pancakes properly, I just gulped them down because I'm always so hungry. Yesterday, for instance, she gave me no dinner. 'Why feed you, imbecile?' she asks.

However [*looks at his watch*], we've somewhat erred and strayed from our subject. So let's go on, though I've no doubt you'd rather hear a song, a symphony or an aria or something. [*Sings.*] 'We'll not be daunted in the heat of battle.' I don't remember where that comes from. By the way, I forgot to say that besides being matron in the wife's school of music, I also have the job of teaching mathematics, physics, chemistry, geography, history, singing scales, literature and all that. The wife charges extra for dancing, singing and drawing, though I'm also the singing and dancing master. Our school of music is at Number Thirteen, Five Dogs Lane. That's probably why I've always had such bad luck—living at Number Thirteen. My daughters were born on the thirteenth of the month too, and the house has thirteen windows. But what's the use of talking? The wife's available at home to interview parents at any time, and if you want a prospectus, they're on sale in the porter's lodge at thirty copecks each. [*Takes several prospectuses from his pocket.*] Or I can let you have some of these if you like. Thirty copecks a whack. Any takers? [*Pause.*] None? All right then, I'll make it twenty. [*Pause.*] How annoying. That's it, house Number Thirteen.

I'm a complete failure, I've grown old and stupid. Here am I lecturing and looking pretty pleased with myself, when I really feel like screaming or taking off for the ends of the earth. There's no one to complain to, it's enough to bring tears to your eyes. You'll say I have my daughters. What of my daughters? They only laugh when I talk to them. My wife has seven daughters. No, sorry—six, I think. [*Eagerly.*] It's seven! Anna, the eldest, is twenty-seven, and the youngest is seventeen. Gentlemen! [*Looks around him.*] Down on my luck I may be, pathetic and foolish I may have become, but in fact you see before you the happiest of fathers—I've no choice in the matter actually, I don't dare say anything else. But if only you knew! Thirty-three years I've lived with my wife—the best years of my life, I might say. Or then again I might not. But in fact they've flashed past like a single moment of ecstasy actually, God damn and blast them! [*Looks round.*]

Anyway, I don't think she's turned up yet—she isn't here, so I can say what I like. I really get the willies when she looks at me. Anyway, as I was saying—the reason why my daughters have been so long finding husbands is probably that they're shy and never meet any men. The wife won't give parties and never has anyone in to a meal. She's a very stingy, bad-tempered bitchy kind of a lady, so no one ever comes to see us, but, er, I can tell you in confidence—. [*Approaches the footlights.*] My wife's daughters are on view on high days and holidays at their Aunt Natalya's—that's the one who has rheumatism and goes round in a yellow dress with black spots on, looking as if she had black-beetles all over her. Snacks are served too, and when my wife's away you can get a bit of you know what. [*Makes a suitable gesture to indicate drinking.*]

One glass is enough to make me drunk, I might add. It feels good, but indescribably sad at the same time. Somehow the days of my youth come back to me, I somehow long—more than you can possibly imagine—to escape. [*Carried away.*] To run away, leave everything behind and run away without a backward glance. Where to? Who cares? If only I could escape from this rotten, vulgar, tawdry existence that's turned me into a pathetic old clown and imbecile! Escape from this stupid, petty, vicious, nasty, spiteful, mean old cow of a wife who's made my life a misery for thirty-three years! Escape from the music, the kitchen, my wife's money and all these vulgar trivialities! Oh, to stop somewhere in the depths of the country and just stand there like a tree or a post or a scarecrow on some vegetable plot under the broad sky, and watch the quiet, bright moon above you all night long and forget, forget! How I'd love to lose my memory! How I'd love to tear off this rotten old tail-coat that I got married in thirty years ago [*tears off his tail-coat*], the one I always wear when I lecture for charity. So much for you! [*Stamps on the coat.*] Take that! I'm a poor, pathetic old man like this waistcoat with its shabby, moth-eaten back. [*Shows the back.*] I don't need anything. I'm above all these low, dirty things. Once I was young and clever and went to college. I had dreams and I felt like a human being. Now I want nothing—nothing but a bit of peace and quiet. [*Glancing to one side, quickly puts on his tail-coat.*]

I say, my wife's out there in the wings. She's turned up and she's waiting for me there. [*Looks at his watch.*] Time's up. If she asks, please, please tell her the lecture was, er—that the imbecile, meaning me, behaved with dignity. [*Looks to one side and coughs.*] She's looking this way.

[*Raising his voice.*] On the supposition that tobacco contains the terrible poison to which I have just alluded, smoking should on no account be indulged in. I shall therefore venture to hope, in a manner of speaking, that some benefit may accrue from this lecture on 'Smoking is Bad for you'.

That's the end. *Dixi et animam levavi!*

[*Bows and struts out majestically.*]

THE NIGHT BEFORE THE TRIAL

[*Ночъ перед судом*]

(the 1890s)

CHARACTERS

FRED GUSEV, a gentleman of advanced years

ZINA, his young wife

ALEXIS ZAYTSEV, a traveller

The keeper of a coaching inn

The action takes place inside a coaching inn on a winter's night. A gloomy room with soot-blackened walls. Large sofas upholstered in oilcloth. A cast-iron stove with a chimney stretching across the whole room.

ZAYTSEV, *with a suitcase; the* INNKEEPER, *with a candle.*

ZAYTSEV. I say, what an awful stink, my good man! You can't breathe in here—there's a sort of sour smell mixed up with sealing-wax and bed-bugs. Ugh!

INNKEEPER. You can't help smells.

ZAYTSEV. Give me a call at six tomorrow, will you? And I want my carriage ready then, I'm due in town by nine.

INNKEEPER. All right.

ZAYTSEV. What time is it now?

INNKEEPER. Half past one. [*Goes out.*]

ZAYTSEV [*taking off his fur coat and felt boots*]. It's freezing! My brains have congealed, I'm so cold. I feel as if I'd been plastered with snow and drenched with water—and then flogged within an inch of my life! What with these snow-drifts and this hellish blizzard, another five minutes out there would have done for me, I reckon. I'm dead beat. And what's it all in aid of? I wouldn't mind if I was on my way to meet a girl or pick up some money I'd inherited—but in fact, you know, I'm on the road to ruin. What a frightful thought! The assizes meet in town tomorrow, and I'm headed for the dock on charges of attempted bigamy, forging my grandmother's will to the tune of not more than three hundred roubles, and the attempted murder of a billiards-marker. The jury will see I'm sent down, no doubt about it. It's here today, in jug tomorrow—and Siberia's frozen wastes in six months' time. Brrr! [*Pause.*] There is a way out of this mess, though, oh yes there is! If the jury finds against me, I'll appeal to an old and trusty friend. Dear, loyal old pal! [*Gets a large pistol out of his suitcase.*] This is him! What a boy! I swapped him with Cheprakov for a couple of hounds. Isn't he lovely! Why, just shooting yourself with this would be a kind of enjoyment. [*Tenderly.*] Are you loaded, boy? [*In a reedy voice, as if answering for the pistol.*] I am that. [*In his own voice.*] I bet you'll go off with a bang—one hell of a ruddy great bang! [*In a reedy voice.*] One hell of a ruddy great bang! [*In his own voice.*] Ah, you dear, silly old thing. Well, lie down and go to sleep. [*Kisses the pistol and puts it in his suitcase.*] The moment they bring in that 'guilty' verdict, I'll put a bullet in my brains and that will be that. I say, I'm frozen stiff. Brrr! Must warm up. [*Does some jerks with his arms and jumps about near the stove.*] Brrr!

[ZINA *looks through the door and immediately disappears from view.*]

ZAYTSEV. What's that? Didn't someone look through the door just then? Well, now! Yes, so they did. So I have neighbours, eh? [*Listens by the door.*] Can't hear a thing, not a sound. Must be some other travellers. How about waking them and making up a four at bridge if they're reasonable people? Grand slam in no trumps! It might help to pass the time, so help me! Even better if it's a woman! Frankly, there's nothing I like more than a wayside romance. Travelling around, you sometimes run into an affair better than any in Turgenev's novels. I remember a situation like this when I was on my travels down Samara way. I put up at a post-house. It's night, see, with the cricket chirping away in the old stove and not another sound. I'm sitting at the table drinking tea, when suddenly there's this mysterious rustle. The door opens and——

ZINA [*from the other side of the door*]. This is maddening, it's grotesque! Call this a post-station! It's an absolute disgrace! [*Looking through the door, shouts.*] Landlord! Where's the landlord? Where are you?

ZAYTSEV [*aside*]. Lovely creature! [*To her.*] The landlord's not here, madam. The lout's asleep. What do you want? Can I be of service to you?

ZINA. It's horrible, ghastly! I think the bed-bugs must want to eat me alive.

ZAYTSEV. Really? Bed-bugs? I say, what awful cheek!

ZINA [*through tears*]. It's quite horrible, in fact. I'm leaving at once. Tell that scoundrel of an innkeeper to harness the horses. Those bugs have really done for me.

ZAYTSEV. You poor girl. To be so lovely and suddenly—. No, this is quite monstrous.

ZINA [*shouts*]. Landlord!

ZAYTSEV. Young lady, er, miss——

ZINA. Not miss—Mrs.

ZAYTSEV. So much the better. [*Aside.*] What a honey! [*To her.*] Madam, what I'm trying to say is this: not being privileged to know your name, and being for my part a gentleman and man of honour, I venture to offer my services. Let me aid you in distress.

ZINA. But how?

ZAYTSEV. I have an excellent habit—always carry insect-powder on my travels. May I offer you some sincerely, from the bottom of my heart?

ZINA. Oh, thank you very much.

ZAYTSEV. Then I'll give it you at once, this very instant—I'll get it out of my suitcase. [*Runs to his suitcase and rummages.*] Those eyes, that nose! This means an affair, I feel it in my bones! [*Rubbing his hands.*] I have all the luck—I've only to fetch up in some wayside inn to have a little romance. She's so lovely, she's even got *my* eyes flashing sparks! Here we are! [*Goes back to the door.*] Here he is, your friend in need!

[ZINA *stretches her arm out from behind the door.*]

ZAYTSEV. No, please—let me come to your room and put it down myself.

ZINA. Certainly not—ask you into my room! Whatever next!

ZAYTSEV. Now why not? There's nothing wrong in it, especially—especially as I'm a doctor. From their doctors and hairdressers ladies have no secrets.

ZINA. You really mean you're a doctor, you're not making it up?

ZAYTSEV. On my word of honour.

ZINA. Oh well, in that case carry on. But why should you put yourself out? I can send my husband to you. Fred! Fred, wake up, you great lump!

GUSEV [*off-stage*]. Eh?

ZINA. Come here. The doctor's very kindly offered us some insect-powder. [*Disappears.*]

ZAYTSEV. Fred! Now that *is* a surprise! Most grateful, I'm sure. I need this Fred like a hole in the head, damn and blast him! Barely have I picked the girl up and had the bright idea of pretending to be a doctor, when up pops friend Fred! Talk about pouring cold water on things! I've a good mind not to give her any insect-powder. She's nothing to write home about, either, with that nondescript little face—two a penny, her sort are. Can't stand that kind of woman!

GUSEV [*in a dressing-gown and nightcap*]. How do you do, Doctor? My wife tells me you have some insect-powder.

ZAYTSEV [*rudely*]. That is so.

GUSEV. Please lend us a bit, we're eyebrow-deep in bugs.

ZAYTSEV. Take it.

GUSEV. Thank you very much indeed, most grateful to you. Did you get caught in the blizzard too?

ZAYTSEV. Yes.

GUSEV. Quite so. What awful weather! And where might you be making for?

ZAYTSEV. Town.

GUSEV. That's where we're going too. I've some hard work ahead of me in town tomorrow and I need my sleep, but there's this bed-bug business and I just can't cope. What with bugs, black-beetles and other creepy-crawlies, our coaching stations are revolting places. I know what I'd do if I had my way with these bug-ridden inn-keepers—run 'em in under Article 112 of the Local Magistrates' Penal Code for 'not keeping domestic beasts under proper control'. Most grateful to you, Doctor. Do you specialize in any particular field?

ZAYTSEV. Chest and, er, head complaints.

GUSEV. Quite so. Very pleased to have met you. [*Goes out.*]

ZAYTSEV [*alone*]. What a ghastly old frump! I'd bury him alive in insect-powder if I had my way. I'd like to beat the swine at cards and clean him out good and proper a dozen times over. Better still, I'd play him at billiards and accidentally fetch him one with a cue that would make him remember me for a whole week. With that blob of a nose, those blue veins all over his face and that wart on his forehead, he—he has the nerve to be married to a woman like that! What right has he? It's disgusting! Yes, it's a dirty trick, I must say. And then people ask why I take such a jaundiced view of things. But how can you help being pessimistic under these conditions?

GUSEV [*in the doorway*]. Don't be shy, Zina. He is a doctor, you know. Don't stand on ceremony, ask him. There's nothing to be scared of. If Shervetsov didn't help you, perhaps he will. [*To* ZAYTSEV.] Sorry to trouble you, Doctor, but could you please tell me why my wife has this constricted feeling in her chest? She has a cough, you know, and it feels tight as if she had some sort of congestion. What's the reason?

ZAYTSEV. That would take time. I can't do a diagnosis just like that.

GUSEV. Oh, never mind that—there's time enough. We can't sleep anyway. Look her over, my dear fellow.

ZAYTSEV [*aside*]. Talk about asking for trouble!

GUSEV [*shouts*]. Zina! Oh really, you are silly. [*To him.*] She's shy—quite the blushing violet, same as me. Modesty's all very well in its way, but why overdo things? What—stand on ceremony with your doctor when you're ill? That really is the limit.

ZINA [*comes in*]. Really, I'm hot and cold all over!

GUSEV. Now, that's quite enough of that. [*To him.*] She's been seeing Dr. Shervetsov, I might add. He's a good fellow, very nice chap, always the life and soul of the party, and he knows his stuff too. But—I don't know, I don't trust him! Somehow I don't like the cut of his jib. Now I can see you're not in the mood, Doctor, but do oblige us, please.

ZAYTSEV. I—I, it's all right. I don't mind. [*Aside.*] This beats everything!

GUSEV. You examine her while I go and tell mine host to put the good old samovar on. [*Goes out.*]

ZAYTSEV. Sit down, please. [*They sit.*] How old are you?

ZINA. Twenty-two.

ZAYTSEV. I see. A dangerous age. May I take your pulse? [*Takes it.*] I see. Quite so. [*Pause.*] What are you laughing at?

ZINA. You really are a doctor, I suppose?

ZAYTSEV. I say, whatever next! Who do you take me for? H'm—. Your pulse is all right. Very much so. A nice, plump little pulse—. I adore travelling adventures, damn it. You go on and on and on—then you suddenly meet a little, er, pulse like this. Do you love medicine?

ZINA. Yes.

ZAYTSEV. Now, isn't that nice! Terribly nice, it is! Let me take your pulse.

ZINA. But, but, but—. Don't go too far!

ZAYTSEV. That lovely little voice, those delightful rolling little eyes! One smile's enough to drive a man crazy. Is your husband jealous? Very jealous? Give me your hand. Only let me take your pulse and I'll die of happiness.

ZINA. Now look here, sir, I can see what's in your mind, but I'm not that kind of girl, sir. I'm a married woman and my husband has a position to keep up.

ZAYTSEV. Yes, yes, I know—but can I help it if you're so lovely?

ZINA. I won't permit liberties, sir—. Kindly leave me alone, or I shall have to take steps. I love and respect my husband too much to let some travelling cad make cheap remarks to me. You're quite wrong if you think I—. My husband's coming now, I think. Yes, yes, he's coming. Why don't you speak? What are you waiting for? Come on, then—kiss me, can't you?

ZAYTSEV. Darling! [*Kisses her.*] You pretty little poppet! [*Kisses her.*]

ZINA. Well, well, well——

ZAYTSEV. My little kitten! [*Kisses her.*] You dear little bit of fluff! [*Seeing* GUSEV *coming in.*] One more question—when do you cough more, on Tuesdays or on Thursdays?

ZINA. On Saturdays.

ZAYTSEV. I see. Let me take your pulse.

GUSEV [*aside*]. It looks as if there's been kissing—it's the Shervetsov business all over again. I can't make any sense of medicine. [*To his wife.*] Do be serious, Zina—you can't go on like this, you can't neglect your health. You must listen carefully to what the doctor tells you. Medicine's making great strides these days, great strides.

ZAYTSEV. Quite so. Now, what I have to say is this. Your wife's in no danger as yet, but if she doesn't have proper treatment she may end up badly with a heart attack and inflammation of the brain.

GUSEV. You see, Zina, you see! The trouble you cause me, oh, I'm so upset.

ZAYTSEV. I'll write a prescription at once. [*Tears a sheet of paper out of the register, sits down and writes.*] *Sic transit* . . . two drams. *Gloria mundi* . . . one ounce. *Aquae dest—* . . . two grains. Now, you take these powders, three a day.

GUSEV. In water or wine?

ZAYTSEV. Water.

GUSEV. Boiled?

ZAYTSEV. Boiled.

GUSEV. I'm really most grateful to you, Doctor——

[UNFINISHED]

APPENDIX I

ON THE HIGH ROAD

1. Composition
2. Text

1. COMPOSITION

The play is a dramatized adaptation of Chekhov's short story *In Autumn* [*Осенью*, 1883]. It was presented for censorship on 29 May 1885, but rejected on 20 September of the same year as unsuitable for performance, and remained unpublished during Chekhov's lifetime.

The report of the 'censor of dramatic works', Keyzer von Nilkheim, reads as follows: 'The action takes place at night in an inn on a high road. Among various tramps and scoundrels, who have come into the inn for a warm and a night's lodging, is a gentleman who has gone to the bad and begs the barman to give him a drink on credit. It turns out in conversation that the gentleman took to drowning his sorrows after his wife had deserted him on their wedding day. In search of shelter from the bad weather a lady, in whom the unhappy drunk recognizes his faithless wife, happens to come into the inn. One of the customers brandishes an axe at her out of sympathy with the drunken squire, and the dramatic sketch ends with this attempted murder. In my opinion this gloomy, sordid play cannot be approved for performance.'

The report is countersigned on a top corner: 'Ban. K. Feoktistov, 20 September 1885.'

2. TEXT

The present translation is made from the text in *Works*, 1944–51, vol. xi, itself based on the manuscript copy of the play preserved in the Central Library of the Russian Drama in Leningrad.

APPENDIX II

SWAN SONG

1. Composition
2. Text
3. Variants

1. COMPOSITION

The play is a dramatized adaptation of Chekhov's short story *Calchas* [*Калхас*, 1886]. The play had been completed by 14 January 1887, on which date Chekhov wrote to M. V. Kiseleva: 'I've written a play on four sides of quarto. It will take fifteen to twenty minutes to act. It's the shortest play on earth. The famous actor Davydov [V. N. Davydov (1849–1925)], now playing at Korsh's Theatre, will act in it. It's being published in *The Season*, so it will circulate all over the place. It's much better, generally speaking, to write short pieces than long; they're less pretentious and they come off all right.... What more can one ask? I wrote the play in an hour and five minutes.'

On 19 February 1888 *Swan Song* received its first performance at Korsh's Theatre in Moscow, the most important of the 'private' (as opposed to the 'Imperial') theatres in Russia.

On 17 October 1888 Chekhov wrote in the following terms to A. N. Pleshcheyev, asking that the play should be approved by the Theatrical and Literary Committee of the Imperial Theatres: 'The play has no merit, I attach no importance to it at all, but the fact is that Lensky [A. P. Lensky (1847–1908), the well-known actor and producer] is absolutely set on playing it on a small stage. It's not so much my request as Lensky's.'

The play was duly approved by the Committee, and a copy was transmitted to Lensky on 26 October. Chekhov wrote to Lensky that he had changed the original title, *Calchas*, to *Swan Song*: 'a long [in Russian], sweet-sour name, but I just couldn't think of any other, though I spent a long time trying' (letter of 26 Oct. 1888).

2. TEXT

The present translation is made from the text in *Works*, 1944–51, vol. xi, itself based on that in Chekhov's *Collected Works*, 2nd edition (1902), vol. vii. The text was first published in this form in the collection of Chekhov's *Plays* (St. Petersburg, 1897).

There are three earlier recensions:

(*a*) the '*Season* text'—that published in *The Season* [*Sezon*], an 'illustrated artistic collection' edited by N. P. Kicheyev, no. 1, Moscow, 1887, entitled *Calchas: a Dramatic Study in One Act*;

(*b*) the 'Censor's copy', with the same title, stamped as approved by the censorship on 9 November 1888;

(*c*) 'Rassokhina's text'—that published in lithographed form by Ye. N. Rassokhina's Moscow Theatrical Library, Moscow, 1888, with the title *Swan Song* (*Calchas*).

3. VARIANTS

The '*Season* text' is considerably shorter than that of the later recensions, and ends with Nikita's speech (see p. 42): 'Mr. Svetlovidov, sir . . .' to '. . . Peter! George!' to which were appended the words: 'Is there anyone here? God, the candle's going out.'

The 'Censor's copy' presents an expanded text which continues as in the final version up to the passage from Shakespeare's *Othello*: 'Farewell the tranquil mind . . .' to '. . . circumstance of glorious war!' (see p. 45). This then continues for a further three lines (from 'And, O you mortal engines . . .' to '. . . Farewell! Othello's occupation's gone!' (see *Othello*, Act Three, Scene iii). The 'Censor's copy' also offers an alternative to this speech: another speech from *Othello*, that beginning:

'Had it pleas'd heaven
To try me with affliction . . .' (see *Othello*, Act Four, Scene ii).

'Rassokhina's text' reproduces that of the 'Censor's copy', using the first-mentioned of the two speeches from *Othello* there admitted as alternatives (i.e. the one retained in the final version, but with three extra lines appended at the end).

In the '*Plays*, 1897 text', *Swan Song* assumed its final version with the removal of the three lines beginning 'And, O you mortal engines . . .' at the end of the speech from *Othello*, and the substitution for these of the last two speeches: those of Nikita and Svetlovidov (see p. 45). At the same time, ten years were added to Svetlovidov's age, originally given as fifty-eight, while his stage career was now said to have lasted 'forty-five years' (see p. 41) instead of the original 'thirty-five'.

APPENDIX III

THE BEAR

1. Composition and first performance
2. Text
3. Variants

1. COMPOSITION AND FIRST PERFORMANCE

The Bear was written in February 1888: 'Being at a loose end, I have written a trivial little farce in the French style called *The Bear*' (letter to Ya. P. Polonsky, 22 Feb. 1888). After first publication in *New Time* [*Novoye vremya*] on 30 August, the play was submitted to the dramatic censorship and received an adverse report: 'The unfavourable impression produced by this highly peculiar theme is increased by the coarseness and impropriety of the tone throughout the play, so that I would have thought it quite unsuitable for performance on the stage' (*Chekhov i teatr*, p. 390). This report was overruled in the censor's eventual resolution, which deplored the 'triviality' of the theme but permitted the play to be performed, while recommending that certain 'rude expressions' should be cut out.

On 11 or 12 October, Chekhov wrote to A. S. Suvorin: '*The Bear* has been passed by the censorship . . . and will be put on at Korsh's Theatre. Solovtsov is dying to play in it' (reference being to the actor N. N. Solovtsov, 1856–1902, to whom Chekhov dedicated the play).

The first performance duly took place at Korsh's Theatre on 28 October 1888, and Chekhov was able to report that: 'Solovtsov played phenomenally . . . the audience never stopped roaring with laughter, and the soliloquies were interrupted by applause. Both actors and author had curtain calls at the first and second performances' (letter to I. L. Leontyev, 2 Nov. 1888).

2. TEXT

The present translation is made from the text in *Works*, 1944–51, vol. xi, itself based on that in Chekhov's *Collected Works*, 2nd edition (1902), vol. vii. The text was first published in this form in the 1st edition of the same (1901).

There are three earlier recensions:

(*a*) the '*New Time* text'—that published in the newspaper *New Time* [*Novoye vremya*], 30 August 1888;

(*b*) 'Rassokhina's text'—that published in lithographed form by Ye. N. Rassokhina's Moscow Theatrical Library, October 1888;

(*c*) the '*Plays*, 1897 text'—that published in the collection of Chekhov's *Plays* (St. Petersburg, 1897).

3. VARIANTS

The '*New Time* text' contains a few minor features which appear in altered form in 'Rassokhina's text':

In Scene v, in place of the words of the final version, 'I'll give you something to remember me by!' (see p. 55), the '*New Time* text' has two lines of scurrilous abuse. These are, in literal translation: 'I'll show you Sidor's nanny-goat and Kuzka's mother, may you get a carriage-shaft, a hundred devils and one witch in your mouth'—or, in somewhat freer translation: 'I'll show you where you get off.'

In Scene viii, in addition to the other items characterizing women (see p. 58), the '*New Time* text' says that they are (in literal translation) 'as wayward as cats and as cowardly as hares'—meaning 'no damn good'.

In Scene ix, the '*New Time* text' contains an additional insult after 'Think you own the place, don't you!' (see p. 61): literally, 'you rough beam'—or, in freer translation, 'you clumsy lout'.

In Scene x, after 'Disgraceful! Scandalous!' (see p. 64), 'Rassokhina's text' contains the sentence: 'I'm in such a filthy state right now, you just can't imagine.' This was cut out in the revision leading to the '*Plays*, 1897 text', where the number of scenes was also expanded, from five to eleven, by turning the soliloquies into separate scenes.

In revising the '*Plays*, 1897 text' for the final version, Chekhov made two minor changes which do not call for comment in translation.

APPENDIX IV

THE PROPOSAL

1. COMPOSITION AND EARLY PERFORMANCES

The first reference to *The Proposal* occurs in a letter of 27 October 1888 from Chekhov to A. S. Suvorin: 'I've written another farce, with two male parts and one female.' On 7 November 1888 Chekhov wrote to I. L. Leontyev: 'I've scribbled a lousy little farce specially for the provinces and sent it to the dear old censor. It's a wretched, vulgar, boring little skit, but it will go down well in the provinces. . . . I shan't put it on in St. Petersburg.'

The censor's permission was received on 10 November, but Chekhov changed his decision not to have *The Proposal* staged in St. Petersburg, and gave Leontyev permission to put on a performance by the Artistic Circle of the Capital: 'I give you *carte blanche*. Make of my notoriously stupid play what you like; roll cigarettes out of it for all I care' (letter to I. L. Leontyev, 11 Mar. 1889).

Thus Leontyev became responsible for the première of *The Proposal* in St. Petersburg, which appears to have taken place on 13 April 1889. The play became popular and one especially important early production was that of the Moscow Little Theatre, where it was first performed on 20 February 1891.

2. TEXT

The present translation is made from the text in *Works*, 1944–51, vol. xi, itself based on that in Chekhov's *Collected Works*, 2nd edition (1902), vol. vii. The text was first published in this form in the collection of Chekhov's *Plays* (St. Petersburg, 1897).

There are two earlier recensions:

(*a*) 'Rassokhina's text'—that preserved in manuscript form in the Central Library of the Russian Drama in Leningrad and marked 'passed for performance, 10 November 1888', and also published in lithographed form by Ye. N. Rassokhina's Moscow Theatrical Library, being stamped as 'passed by the censorship, 27 December 1888';

(*b*) the '*New Time* text'—that printed in the newspaper *New Time* [*Novoye vremya*], 3 May 1889.

3. VARIANTS

In revising 'Rassokhina's text' for the '*New Time* text', Chekhov made a number of small alterations.

In Scene iv, after Chubukov's words 'go ahead and sue, sir!' (see p. 75), the '*New Time* text' was expanded to include: 'Oh, I know your sort! Just what you're angling for and so on, isn't it—a court case, what?'

The other changes were all cuts, including the removal of repeated requests from Chubukov to Natasha to 'shut up!' and also the following other material, which is present in 'Rassokhina's text' but not in later recensions:

after the words	*'Rassokhina's text' has*
It was *not*, it was yours! (p. 77)	You're ill-bred and rude. If it wasn't for you he wouldn't have gone away.
So now it's my fault, what? (p. 77)	Just wait, my good woman, and so on and so forth. If I shoot or hang myself, it'll be your fault, I can tell you! It's you! You brought me to this pass!
Now let's change the subject. (p. 77)	Have you been to the fair at Nikitovka?
. . . Rover or Tracker? (p. 79)	CHUBUKOV. Arguing again, are you? I can't stand that.
Why, you young puppy! (p. 80)	Walking chemist's shop!
Nasty old fogy! Canting hypocrite! (p. 80)	I know you only too well!
. . . idle gasbag! (p. 80)	NATASHA. Rover's a million times better than your Tracker. Oxpen Field is ours! So there! CHUBUKOV. Oxpen Field is ours!

In the final text, as first printed in '*Plays*, 1897', Lomov's soliloquy 'I feel cold, I'm shaking like a leaf...' (p. 70) is made into a separate scene (Scene ii), with the result that there were now, in the final version, seven scenes altogether, in place of six in the earlier recensions.

APPENDIX V

TATYANA REPIN

1. Composition
2. Text

1. COMPOSITION

Chekhov's *Tatyana Repin* was not intended for production, but was a sort of private joke between Chekhov and A. S. Suvorin (1834–1912), his closest friend and publisher, who was himself a playwright. Chekhov's *Tatyana Repin* is, in fact, a whimsical continuation of Suvorin's own play with the same title (1886). This in turn was based on a real-life drama concerning the well-known actress Eulalia Kadmina. In 1881 Kadmina appeared in Kharkov in the historical drama *Vasilisa Meletyev* by Alexander Ostrovsky and S. A. Gedeonov, in which her part, as wife of Ivan the Terrible, called for her to be poisoned on the stage. She lent additional realism to the proceedings by taking actual poison and publicly expiring in agonies the more realistic because they were unfeigned: all this to punish a faithless lover who was a member of the audience. The episode inspired the plot of Suvorin's play, where the heroine, Tatyana Repin, commits suicide by poison after her faithless lover, Sabinin, has fallen in love with Vera Olenin. These characters all feature in Chekhov's whimsical continuation, as do various other figures from among Suvorin's *dramatis personae*: the landowners Kotelnikov and Patronnikov; the financier Sonnenstein and others.

Suvorin arranged for Chekhov's play to be printed on his own printing-press, and in two copies only, with the title: '*Tatyana Repin: a One-Act Drama.* By Anton Chekhov (St. Petersburg, 1889).'

After finishing the play, Chekhov wrote to Suvorin as follows (5 Mar. 1889): 'In return for the dictionaries [sent to him by Suvorin] I'm sending you a very cheap and useless present, but one which I alone can give. Wait and see.'

On the following day Chekhov wrote again: 'I enclose . . . the very cheap and useless present which I promised. I shall be bored by your dictionaries, so you may as well be bored by my present. I wrote it at a sitting in a hurry, so it's come out very cheap and vulgar. As for me using your title, you'd better take me to court. Don't show it to anyone, and throw it in the fire when you've finished it. Or else throw it in the fire unread' (letter to A. S. Suvorin, 6 Mar. 1889).

On 14 May Chekhov wrote to Suvorin: 'Thank you, I've received my *Tatyana Repin*. The paper's very good. I crossed out my name in proof, and I can't understand how it came to be left in. I also crossed out (i.e. corrected) a lot of misprints which have survived as well. Anyway, none of that matters. For the sake of greater illusion, Leipzig should have been printed on the cover instead of St. Petersburg.' The 'illusion' consisted of the pretence that the publication was illegal, as in a sense it was: Chekhov's play could not have been passed by the censorship owing to the portrayal of a rite of the church on stage.

2. TEXT

The present translation is made from the text in *Works*, 1944–51, vol. xii, itself based on Chekhov's own copy of the play (one of the two printed by Suvorin), which is preserved in the library of the State Literary Museum in Moscow.

APPENDIX VI

A TRAGIC ROLE

1. Composition
2. Text
3. Variants

1. COMPOSITION

The play is a dramatized adaptation of Chekhov's short story *One Among Many* [*Один из многих*, 1887]. The play was written on 4 May 1889, on which date Chekhov informed A. S. Suvorin: 'Last night I remembered promising Varlamov [the actor K. A. Varlamov (1848–1915)] to write him a farce. I wrote it today, and I've already sent it off.'

On 6 May 1889 Chekhov wrote to I. L. Leontyev: 'A day or two ago I remembered promising Varlamov last winter to turn one of my novels [*sic*] into a play. I sat straight down and did so, making a pretty bad job of it. A novel on a stale and hackneyed theme produced a stale farce that falls flat. It's called *A Tragic Role*.'

2. TEXT

The present translation is made from the text in *Works*, 1944–51, vol. xi, itself based on that in Chekhov's *Collected Works*, 2nd edition (1902), vol. vii. The text was first published in this form in the magazine *Artist* [*Artist*], vol. vii, April 1890.

There is one earlier recension: that published in lithographed form by B. A. Bazarov's Theatrical Library in St. Petersburg, and marked with the date of censorship, 1 June 1889.

3. VARIANTS

In preparing the text for the final version as printed in *Artist*, Chekhov removed two sentences:

after the words	*Bazarov's text has*
. . . so I have to buy him a bicycle. (p. 111)	The Kuridins' baby has died and I've got to take them a coffin.
You have to dance the quadrille yourself. (p. 112)	You dance with some misshapen female and grin like an idiot.

APPENDIX VII

THE WEDDING

1. Composition
2. Text
3. Variants

1. COMPOSITION

The play is a dramatized adaptation of Chekhov's short story *A Wedding with a General* [*Свадьба с генералом*, 1884], but also contains themes from two other stories: *The Wedding Season* [*Свадебный сезон*, 1881] and *Marrying for Money* [*Брак по расчету*, 1884]. The play was written in October 1889 and submitted to the censorship on the 31st of that month, permission to perform it being granted two days later. In April of the following year Chekhov made certain cuts (see below under *Variants*) and once more obtained the censor's approval, dated 25 April 1890.

2. TEXT

The present translation is made from the text in *Works*, 1944–51, vol. xi, itself based on that in Chekhov's *Collected Works*, 2nd edition (1902), vol. vii. The text was first published in this form in a collection of Chekhov's plays, *The Wedding, The Anniversary, Three Sisters* (St. Petersburg, 1902).

There are two previous recensions:

(*a*) the 'Leningrad manuscript'—that preserved in the Central Library of the Russian Drama in Leningrad and marked as 'passed for performance by the censorship, 2 November 1889';

(*b*) 'Rassokhin's text'—that preserved in manuscript form in the State (Lenin) Library of the U.S.S.R. in Moscow, and also published in lithographed form by S. F. Rassokhin in 1890, being marked as 'passed by the censorship, Moscow, 25 April 1890'.

3. VARIANTS

The 'Leningrad manuscript' contains a fair amount of material which was later cut.

In the first stage of revision (leading to 'Rassokhin's text') the following extensive cut was made, the passage in question appearing in the 'Leningrad manuscript' after Yat's words '. . . a weakness in the psychology department, that's all' (see p. 125):

MOZGOVOY [*standing up*]. Ladies and gentlemen, may I say a word? Mrs. Zmeyukin's one down. When we were playing forfeits, her forfeit was to kiss the man with the darkest hair. Well, Lapkin had the darkest hair, but he came over all shy and went away. Now Mrs. Zmeyukin really should pick some other man with dark hair and pay her forfeit.

ALL. Yes, yes. Of course, of course.

MRS. ZMEYUKIN. Oh, rubbish—leave me alone.

MOZGOVOY. Who has the darkest hair?

BEST MAN. Mr. Dymba.

MRS. ZMEYUKIN. I don't want to kiss anyone. [*Covers her face with a napkin.*]

YAT. Now, now, Mrs. Zmeyukin, that won't do—don't you try and wriggle out of it. You come here.

DYMBA [*embarrassed*]. Why, pliss? I don't mind, isn't it——

MRS. ZMEYUKIN. I really can't see—. This is rather odd. [*Aside.*] He's so handsome. What eyes!

YAT [*rubbing his hands*]. No, no, that won't do. You've got to.

[*The band plays a flourish.*]

MRS. ZMEYUKIN. All right then. If that's my forfeit, very well, but—it's overdoing things a bit.

[*Kisses* DYMBA. *The kiss is a prolonged one.* DYMBA *retreats and balances with his arms as* MRS. ZMEYUKIN *fastens her lips on his.*]

YAT [*alarmed*]. It's on the long side. That's enough. Cut it out.

ALL [*alarmed*]. Enough! Stop it!

DYMBA [*wrenching himself away from* MRS. ZMEYUKIN]. Ugh! Why, pliss? I don't understand, isn't it?

YAT, MOZGOVOY [*together*]. She's going to faint! Fetch water! She's passing out!

YAT [*giving water to* MRS. ZMEYUKIN, *aside*]. Quite the lady of the manor, ain't she? [*To her.*] Calm down. Please! That's better, she's opened her eyes. [*Aside.*] How frightfully la-di-da!

MRS. ZMEYUKIN [*coming to*]. Where am I? What atmosphere is this? Where is he? [*To* DYMBA.] You devil! You set me ablaze with that kiss!

DYMBA [*embarrassed*]. Why, pliss, isn't it?

The same revision involved three other minor cuts:

after the words	*the 'Leningrad manuscript' has*
Mess-sewers, promenade! (see p. 120)	*Donnez-moi toujours.*
Have you established civil servants too? (see p. 122)	YAT. But I bet Greece hasn't got such beautiful persons of the female sex as some I could mention.

. . . he's a General of the Fleet. (see p. 126)	He gave orders to the seas and commanded the storms.

Further fairly extensive changes were made when Chekhov prepared the text of the final version.

After Mrs. Zhigalov's words '. . . a General you shall have' (see p. 120), the following was cut:

> We'll lay down the red carpet. But if we don't get one, it means Mr. Nyunin couldn't find a suitable specimen. It's time we sat down to supper—it's getting on for midnight.

After Aplombov's words 'Surely you knew I'd find it awkward?' (see p. 120), the following cut was made:

MRS. ZHIGALOV. We never asked him, my dear man. He just turned up.

APLOMBOV. I don't want to make a scene, or else I'd teach him to gatecrash.

Yat's words in the final version, 'I'm in love, head over heels!' (see p. 122), replace the following passage in the earlier drafts:

> She's an ugly, common or garden midwife, but look at the airs she gives herself. Quite the lady of the manor, ain't she?

After Mrs. Zhigalov's words, 'Do sit down, General' (see p. 127), the following cut was made:

> [*Aside.*] But what an unimposing, shop-soiled specimen! Hasn't even got any epaulettes! Well, at least he has a lot of medals. [*To him.*]

In addition to the above changes, Revunov's speeches were very largely recast, so that much of the technical sailor's jargon (see pp. 128–30) appears in different form in the drafts. There seems little advantage in including in the present edition a translation of the variants to this material, which must remain largely a mystery to anyone without sailing experience of a special kind—and in which, incidentally, Chekhov's text makes a number of technical mistakes.

Chekhov attached great importance to the second revision of *The Wedding*, about which he wrote to his publisher A. F. Marks in the following terms: 'My farce *The Wedding* was sent to you [in an earlier draft] two years ago with the note "not to go into the complete edition". Then it was considerably revised by me and sent to you this year. . . . Now I've got the proofs and it turns out that the type was set, not from the corrected copy, but from the old one. Please have it set up from the corrected copy, but if that's been lost we shall have to scrap *The Wedding* as I haven't a spare copy' (letter of 25 Dec. 1901).

APPENDIX VIII

THE ANNIVERSARY

1. Composition
2. Text
3. Variants

1. COMPOSITION

The play is a dramatized adaptation of Chekhov's short story *A Defenceless Creature* [*Беззащитное существо*, 1887]. The first mention of *The Anniversary* occurs in a letter from Chekhov to S. F. Rassokhin of 17 December 1891, in which Chekhov asks Rassokhin to have two copies of the play made and submitted for censorship, which was done on 21 December. Permission to perform the play was granted on 30 December. In February of the following year it was published in lithographed form by Rassokhin. Chekhov revised it in 1901 for inclusion in his *Collected Works*, 2nd edition (1902), vol. vii.

2. TEXT

The present translation is made from the text in *Works*, 1944–51, vol. xi, itself based on that in Chekhov's *Collected Works*, 2nd edition (1902), vol. vii. The text was first published in this form in a collection of Chekhov's plays, *The Wedding, The Anniversary, Three Sisters* (St. Petersburg, 1902).

There are two previous recensions:

(*a*) the 'Bakhrushin manuscript'—a recently discovered manuscript (preserved in the A. A. Bakhrushin Theatrical Museum and with the presumed date Dec. 1891), which was not available to the editors of *Works*, 1944–51, xi, but has since been described in *Chekhov: Literaturnoye nasledstvo*, p. 129;

(*b*) 'Rassokhin's text'—that published in lithographed form by S. F. Rassokhin in February 1892 and marked as passed by the Moscow censor.

3. VARIANTS

The 'Bakhrushin manuscript' was the basis of Rassokhin's edition, but offers one or two minor variants in places where an alteration has been made, but an earlier, excised version still remains legible. The following exchange, occurring after 'Well, I can't see why you like them so' (p. 139), was excised at this stage:

KISTUNOV (= SHIPUCHIN). Why? Well, er—women, old man, are just one of those things. They're a sort of—. They're the spice of life! But you go on writing, my dear fellow. You must hurry.

KHIRIN. There's different kinds of spices. [*Pause.*]

A number of other minor variants in the corrected manuscript are recorded in *Chekhov: Literaturnoye nasledstvo*, p. 129, but do not call for attention in translation.

Comparison of 'Rassokhin's text' with the final version shows that Chekhov made a large number of alterations, chiefly cuts, at this stage. He changed the names of two of the characters, 'Shipuchin' appearing in the earlier drafts as 'Kistunov', and 'Mrs. Merchutkin' as 'Mrs. Shchukin'. He also made the insertion 'or my name's not Shipuchin' in a large number of places. The other changes include one major cut, consisting of the following scene with which the play originally ended:

[TATYANA *lies groaning on the sofa.*]

KHIRIN [*after a long pause*]. What did I say, eh? They come in here, ruin everything and kick up a row! One of them picks up twenty-five roubles and goes away, while the other creature lies there—. [*Points to* TATYANA.] Quite a distinguished record! I've told you hundreds of times, you shouldn't touch 'em with a barge-pole.

KISTUNOV (= SHIPUCHIN). Deputation, reputation—. Old woman, wife, felt boots—. Someone shot himself.

'Two friends one night went for a walk,
And on that walk they had a talk.'

[*Rubbing his eyes.*] I've spent a fortnight making up a shareholders' address, I've bought a silver tankard at my own expense and put down seventy-five roubles of my own money to have the address bound. I've spent five days and nights standing in front of a mirror and seeing that my suit was just right—and now what? It's all gone down the drain, I tell you! We're disgraced, done for! My reputation's ruined.

KHIRIN. And whose fault's that? It's yours, you know—you've made a complete hash of things.

KISTUNOV. Shut up! It's your fault, not mine.

KHIRIN. It's yours, I tell you.

KISTUNOV. No, it's yours. If it wasn't for your beastly felt boots and your impossible bloody character, it would have been all right. Why did you have to chase after my wife? Why shout at her? How dare you?

KHIRIN. Well, if you weren't always chasing women yourself and trying to put one over on people—. But I don't want to work here any more, damn it!

You give me my gold medal and three-hundred-rouble bonus. Come on, hand over.

KISTUNOV. You won't get a thing, old sausage. And you can go and take a running jump at yourself!

KHIRIN. Oh, can I! Then so much for your speech! [*Tears up the speech.*] There you are! I'll show you where you get off! You wait!

KISTUNOV [*shouts*]. Get out of here! [*Rings.*] Hey, you there! Get him out of here.

KHIRIN [*stamps his feet*]. Clear out of my sight. I'll stop at nothing, I don't answer for myself. Out!

KISTUNOV. Go away!

[*They chase each other with shouts of* 'Get out, go away!' *Noise. Clerks run in.*]

CURTAIN

The remaining changes were numerous, but all affected comparatively short passages.

The following passage, originally occurring after 'He only wants to put over a lot of eyewash, and I'm' (see p. 137), was cut:

> to tot up figures, calculate the interest, make a neat copy, and

The following passage, originally occurring after Shipuchin's words 'for the sake of the Bank's reputation' (see p. 139), was cut:

> Only I don't know where the ceremony of reading the address should take place—in the club before dinner or here? I'd like it to be here and I hinted as much to them.

The following passage, originally occurring after 'My great merit is simply that I've raised the Bank's prestige' (see p. 139), was cut:

> Just you listen to what the public thinks. 'It's more like a government department than a bank,' say they. 'We're scared to go inside,' they say.

Where the final version has 'Is it true my husband . . .' to '. . . on the stock exchange?' (see p. 140), the earlier draft has:

> Is it true that the Bank's lent out four hundred thousand with no security?

The following sentence was added after '. . . everyone's angry with you for not writing' (see p. 141):

> Zina sends a kiss too.

From Tatyana's lines (p. 141) the italicized passages were cut:

I'm scared to talk about it, really. *It was a tragedy.* What a business! *If it was printed in the newspapers everyone would be horrified. . . .*

I'm so sorry for her! *Listen carefully, Andrew—you'll be horrified.*

After Tatyana's 'so what's to be done?' (p. 143), the following was cut:

If she marries him, what will they live on? After all, you can't live on love alone.

The following cuts were made in Mrs. Merchutkin's lines.
After '. . . and my son-in-law out of work as well' (p. 144):

I hardly eat and drink at all, and I barely keep body and soul together. I got no sleep all night.

After 'And I can show you the doctor's certificate if you want' (p. 145):

[*In a whisper.*] Do something for me, kind sir, I shan't forget to show I'm grateful. That's right, there's a rouble in this for you.

After 'But what about my money?' (p. 145):

I'm a poor woman, with nothing to live off but my lodgers.

After 'Who do you think you're talking to?' (p. 145):

The General's more of a big-wig than you, but he doesn't shoot his mouth off—and there are you waving your arms about. I wouldn't be in such a hurry to wave those arms about!

After '. . . I know your sort' (p. 145):
If I go and see the lawyer Dmitry Karlovich, you'll end up in Queer Street. I've taken three lodgers to court, and I'll have you grovelling at my feet for your rudeness.

After 'Where was you brought up?' (p. 146):

You wait till the General comes in—I'll settle your hash!

After 'I'm a weak, defenceless woman, I am' (p. 146):

I won't have it. My husband's an established civil servant and I'm the daughter of a lieutenant.

After Khirin's 'This beats everything' (p. 146), the following was cut:

Or leave it to me. I'll deal with her myself.

After Shipuchin's words '. . . there are a lot of private apartments in this block' (p. 146), the following was cut:

and the devil only knows what people may think of us. Our reputation will go down the drain.

After Khirin's words 'I can't stand this' (p. 146), the following was cut:

I won't answer for myself, I'll stop at nothing.

After Shipuchin's words '. . . a private business establishment' (p. 146), the following was cut:

What do you want from us? Can't you get it into your head that you're being a nuisance? [*To his wife.*] Go home, Tatyana dear.

TATYANA. Straight away. I'll finish my business and go

Instead of Mrs. Merchutkin's 'Pretty lady . . .' to '. . . didn't go down well at all, it didn't' (p. 147), the earlier draft has:

I'm an orphan, my pretty darling, there's no one to stick up for me. I might as well be a widow, I'm worse off than any unfortunate widow.

After Khirin's speech to Tatyana 'Beat it!' (p. 148), the earlier draft reads:

TATYANA. What! How dare you!

KISTUNOV (= SHIPUCHIN) [*iumps up*]. Stop it! Are you out of your mind? Be quiet. [*To* MRS. SHCHUKIN (= MERCHUTKIN).] Get out of here!

The above was replaced by the passage 'TATYANA. What! *What* did you say?' to 'SHIPUCHIN. . . . Get out of here, I tell you.'

Instead of 'MRS. MERCHUTKIN [*groans*]. Oh, oh!' (p. 149), the draft has:

MRS. SHCHUKIN (= MERCHUTKIN) [*coming to*]. All right, my good sir. Oh, confound you! I nearly died of fright. I'll be back again tomorrow.

[*Goes out. The shareholders make way for her.*]

APPENDIX IX

SMOKING IS BAD FOR YOU

1. Composition
2. Text
3. Variants

1. COMPOSITION

The question arises as to whether *Smoking is Bad for You* is to be regarded (as in the present edition) as one play, or (as in *Works*, 1944–51, xi) as two different plays, one dated 1889 and the other 1903. The latter solution has the support of Chekhov himself, who wrote as follows on 1 October 1902 to A. F. Marks, the publisher of his first *Collected Works*: 'Among works of mine transmitted to you is the farce *Smoking is Bad for You*, which is one of the items which I asked you to exclude from my *Complete Collected Works* and never print [it was Chekhov's desire that certain *juvenilia* or otherwise unsatisfactory writings should not appear in Marks's edition of his works]. Now I have written a completely new play with the same title, *Smoking is Bad for You*, keeping only the surname of the *dramatis persona*, and I send it you for inclusion in volume vii.'

Marks replied asking Chekhov's permission to publish the play first in the magazine *Meadow* [*Niva*], but Chekhov answered on 16 October that his farce was written solely for the stage, and that it 'might seem pointless and boring in a magazine'. It eventually appeared, not in vol. vii of the *Collected Works*, as envisaged, but in vol. xiv (1903).

Though this version is stated in *Works*, 1944–51, xi, p. 595, to have been 'written in September 1902', it does in fact represent the sixth recension of what can reasonably be regarded as a single play, the first draft having been finished on 14 February 1886. On that date Chekhov informed V. V. Bilibin: 'I've just finished my monologue *Smoking is Bad for You*, which was intended in my heart of hearts for the comic actor Gradov-Sokolov [L. I. Gradov-Sokolov (1845–90)]. Having only two and a half hours at my disposal, I spoilt this monologue and . . . sent it, not to the devil, but to the *Petersburg Gazette*. My intentions were for the best, but the execution turned out quite execrable.'

Though the monologue was thoroughly transformed in the course of five main revisions spread over sixteen years, it still retains, in its final shape, some material from the 1886 version, and Chekhov does not seem justified in regarding his text of 1903 as an 'entirely new work'. He perhaps felt it necessary to emphasize this point to Marks because on at least one other occasion Marks had, through a misunderstanding, set up the type of a superseded earlier version (of *The Wedding*) instead of a revised version approved by Chekhov (see p. 183).

2. TEXT

The present translation is made from the text in *Works*, 1944–51, vol. xi, itself based on Chekhov's *Collected Works*, vol. xiv (1903), where the text was first published in this form.

As already mentioned, there are five earlier recensions:

(*a*) the '*Petersburg Gazette* text'—that published in the *Petersburg Gazette* [*Peterburgskaya gazeta*] on 17 February 1886, signed A. Chekhonte;

(*b*) the '*Motley Stories* text'—that published in the collection of Chekhov's early work, *Motley Stories* [*Pyostryye rasskazy*, St. Petersburg, 1886];

(*c*) 'Napoykin's text'—that published in lithographed form by S. I. Napoykin's Theatrical Library and marked as passed by the censorship, Moscow, 31 January 1887; a second edition of Napoykin's text was brought out in Moscow in 1889 and is identical with

(*d*) 'Rassokhina's text I'—that published in lithographed form by Ye. N. Rassokhina's Moscow Theatrical Library and marked as passed by the censorship, Moscow, 30 May 1889;

(*e*) 'Rassokhina's text II'—being the second edition of the above, dated 1890 and containing, in the copy preserved in S. M. Chekhov's archive, numerous alterations made at an unknown date in Chekhov's hand.

3. VARIANTS

In order to give as full a picture as is feasible of the variants, the following procedure is adopted here:

(i) 'Rassokhina's text I' (1889) is translated in full;

(ii) an account is given of the principal variants presented in the three preceding recensions (of 1886–7);

(iii) 'Rassokhina's text II' (1890, with later alterations of unknown date) is translated in full.

(i) '*Rassokhina's text I*'

SMOKING IS BAD FOR YOU

A MONOLOGUE

(1889)

CHARACTER

MARCELLUS NYUKHIN, a hen-pecked husband whose wife keeps a girls boarding-school

The stage represents the platform in the hall of a provincial club

NYUKHIN *struts in majestically, bows, adjusts his waistcoat and begins majestically.*

NYUKHIN. Ladies and gentlemen.

It's been suggested to the wife that I should lecture here in aid of charity on some topic of general interest. True learning is modest and avoids ostentation, but seeing it's for charity the wife agreed—and so I stand before you now.

I'm not a professor, and university degrees have passed me by, but it's common knowledge that I—that I [*hesitates and glances at a piece of paper which he takes out of his waistcoat pocket*]—have sacrificed my health and creature comforts for these last thirty years working non-stop on problems of a strictly academic nature, and that I even have some of my learned scientific articles printed in the local rag. A few days ago I gave the editor a large article entitled: 'The Ill Effects of Coffee-itis and Tea-mania on the Organism'.

As the subject for my lecture today I've chosen the harmful effects of smoking and taking snuff on the human race. To explore the full significance of the theme in the scope of a single lecture is naturally difficult, but I shall endeavour to be brief and confine myself to essentials. Opposed as I am to the popular approach, I shall be strictly academic, and I suggest that you members of my audience should be imbued with a sense of the subject's significance and attend, with due seriousness, to my present lecture. If anyone's scared by the idea of a dry, strictly scientific speech—if he's that frivolous, he can stop listening and go. [*Makes a majestic gesture and adjusts his waistcoat.*] Very well, I commence.

Your attention, please. I should like to ask the doctors in my audience to pay particular attention. My lecture is a mine of useful information for them, since nicotine not only has harmful effects, but is also used in medicine. For instance, it was prescribed to my wife in the form of an enema on the 10th of February 1871. [*Looks at his piece of paper.*] Tobacco is an organic body. It is derived, to my way of thinking, from the plant *Nicotiana Tabacum*, which belongs to the genus *Solaneae*. It grows in America. Its main constituent component is an 'orrible deadly poison—nicotine. Chemically, to my way of thinking, it consists of ten atoms of carbon, fourteen atoms of hydrogen and two, er, atoms of nitrogen. [*Pants and clutches his chest, dropping his piece of paper.*]

Give me air! [*Balances with his arms and legs to stop himself falling over.*] Whew! Just a moment! Let me get my breath back! Just a moment. One minute. I shall stop this attack by sheer will-power. [*Beats his chest with his fist.*] That will do. Gosh! [*A minute's pause, during which* NYUKHIN *walks up and down the stage, panting.*]

I've suffered from these choking bouts—asthma—for ages. This complaint dates from the 13th of September 1869, the day when my wife gave birth to her sixth daughter, Veronica. The wife has nine girls in all—but no boys, of which she's very glad because boys would be a nuisance in a girls' boarding-school in many ways. There's only one man in the whole school—myself. But the respectable and distinguished families who have confided their children's fate to my wife can rest assured as far as I'm concerned. Anyway—time being short, let's not wander from the subject in hand. Now then, where were we? Phew! That choking fit caught me at the most interesting point. But it's an ill wind that blows nobody any good. For me and you, and especially for the doctors in the house, that attack can serve as an excellent lesson. There are no effects in Nature without a cause. Let us therefore seek the cause of my present choking fit. [*Places a finger on his forehead and thinks.*]

Yes, the only cure for asthma is to avoid heavy and spicy foods, but before coming here to lecture I permitted myself a certain indulgence. It must be added that today was pancake day at my wife's boarding-school. At lunch each girl receives a single pancake instead of the main course. Being my wife's husband, I don't think it's my place to praise a woman of such integrity, but I swear that nowhere is catering so rational, hygienic and efficient as at the wife's school. I can bear witness to this myself, having the honour to be matron. I buy food, keep an eye on the servants, present the accounts to my wife every evening, make up exercise-books, devise anti-insect precautions, spray the air, count the linen, make sure there's at least one tooth-brush per five girls, and that not more than ten of them dry themselves on the same towel. Today I had the job of issuing flour and butter to Cook in a quantity strictly corresponding to the number of the girls. Well, so we had pancakes today. It must be added that the pancakes were intended solely for the girls. For the members of my wife's family a roast was prepared, for which purpose we had a shin of veal that had been kept in the larder since last Friday. The wife and I decided that it might spoil by tomorrow if we didn't cook it today. Anyway, let's go on.

Now, what do you think happens next? When the pancakes are already cooked and counted, the wife sends to the kitchen to say that five girls have been punished for misbehaviour by not being allowed pancakes. It thus transpires that we have five pancakes in hand. What are we to do with them? Quite. Are we to give them to the daughters? But my wife won't let her own girls eat stodgy foods, so what do you think we did with them? [*Sighs and shakes his head.*] Could anything have been kinder, more loving, more angelically good? 'Marcellus dear,' says the wife, 'you eat 'em yourself.' So I ate them, after drinking a preliminary glass of vodka. That's why I got the wheezes, that's what was behind it all. However—. [*Looks at his watch.*]

We've somewhat erred and strayed from our subject. So let's go on. Well, chemically speaking, nicotine consists of, er—[*nervously rummages in his pockets and looks for his piece of paper*]. I suggest you memorize this formula. A chemical formula's a great stand-by. [*Seeing his piece of paper, drops his handkerchief on it. Picks up the paper and handkerchief together.*]

I forgot to say that, besides being matron in the wife's school, I also have the job of teaching mathematics, physics, chemistry, geography, history and visual aids. These subjects apart, my wife's school supplies tuition in French, German, English, Scripture, needlework, drawing, music, dancing and deportment. As you see, it has a larger curriculum than the grammar schools, not to mention the food! And the comfort! And you get all this for practically nothing, that's what's so fantastic! Full board costs only three hundred roubles, half board is two hundred, and day girls pay a hundred. There's an extra charge for dancing, music and drawing by agreement with my wife. It's a fine school! It's located on the corner of Cat Street and Five Dogs Alley in Mrs. Mamashechkin's house—the one whose husband was a major. The wife's at home available to interview parents at any time, and the school's prospectus is on sale in the porter's lodge at fifty copecks a copy. [*Looks at his piece of paper.*]

So I suggest you memorize the formula. Chemically speaking, nicotine consists of ten atoms of carbon, fourteen of hydrogen and two of nitrogen. Kindly make a note of it. It consists of a colourless liquid which smells like ammonia. But what matters to us, actually, is the immediate effect of nicotine [*looks in his snuff-box*] on the nervous centres and muscles of the digestive canal. Oh Lord, they've been mucking round with it again! [*Sneezes.*] Now, what am I to do with these wretched, miserable girls? Yesterday they put face-powder in my snuff-box, today it's something with an acrid stink. [*Sneezes and scratches his nose.*] Sickening! God knows what this stuff's doing to my nose! Ugh! What rotten, nasty little girls! Perhaps you feel that this misdemeanour argues a lack of discipline in the wife's school. No, my dear sirs, it's not the school's fault, indeed no! It's society's fault, it's your fault! Family and school should march hand in hand, but what do we see? [*Sneezes.*] But let's forget this! [*Sneezes.*] Forget it.

Nicotine puts the stomach and intestines in a tetanic condition, that is in a condition of tetanus. [*Pause.*]

But I notice smiles on many faces. Obviously not all members of the audience fully appreciate the supreme importance of our theme. Some people even think it funny when they hear the hallowed austerities of Science proclaimed from the podium. [*Sighs.*] Naturally I don't venture to rebuke you, but—. 'Children,' I always tell my wife's daughters, 'don't laugh at what's no laughing matter!' [*Sneezes.*]

My wife has nine daughters. Anna, the eldest, is twenty-seven, and the youngest is seventeen. Gentlemen! These nine young, unspoilt creatures are an amalgam of everything beautiful, pure and exalted in Nature. Pardon my emotion and the catch in my voice, but you see before you the happiest of fathers. [*Sighs.*] But how difficult it is to get a girl married these days. Terribly hard, it is. It's easier to borrow money by mortgaging your property three times over than it is to find a husband for even one of these daughters.

[*Shakes his head.*] Ah me, young men, young men—by your stubbornness and materialist leanings you deprive yourselves of one of the highest pleasures, that of family life. If only you knew what a good life it is. Thirty-three years I've lived with my wife—the best years of my life, I might say. They've flashed past like a single moment of ecstasy. [*Weeps.*] How often have I distressed her by my weaknesses. My poor wife! Although I've meekly accepted punishment, how have I rewarded her anger? [*Pause.*]

The reason why my wife's daughters have been so long finding husbands is that they're shy and never meet any men. The wife can't give parties and never has anyone in to a meal, but, er, I can tell you in confidence—. [*Approaches the footlights and whispers.*] My daughters are on view on high days and holidays in their Aunt Natalya's house—that's the one who has epilepsy and collects old coins. Snacks are served.

But let's not digress, time being short. I got as far as tetanus. Anyway [*looks at his watch*]—until we meet again! [*Adjusts his waistcoat and struts out majestically.*]

(ii) *Variants preceding 'Rassokhina's text I'*

Three stages of revision must be considered, at each of which cuts predominated over insertions and alterations. These are the stages which preceded the production of the '*Motley Stories* text'; of 'Napoykin's text'; and of 'Rassokhina's text I' respectively.

The following alterations or cuts were made in the first stage of revision:

Instead of the passage 'Naturally I don't venture . . .' to '. . . what's no laughing matter! [*Sneezes.*]' (p. 193), the '*Petersburg Gazette* text' has:

I attribute this laughter to defective upbringing. One mustn't laugh at what is great, beautiful and sacred. Woe unto him who laughs. My wife's daughters never laugh. That's the way they were brought up, and I can die with a clear conscience.

Instead of 'Pardon my emotion . . .' to '. . . the happiest of fathers' (p. 193), the '*Petersburg Gazette* text' has:

None of them's married yet, but you only have to look at them to be sure that they'll make excellent wives.

After '. . . they're shy and never meet any men' (p. 194), the following passage (cut out of later recensions) occurs in the '*Petersburg Gazette* text':

You ought to have a look at them, young men. Who knows, you might take a fancy to one of the nine.

In the second and third stages of revision—those leading to 'Napoykin's text' and 'Rassokhina's text I' respectively—further cuts were made. These include the following, of which the second part (after 'Moleschott') was removed in stage two and the first part was removed in stage three. It occurred after '. . . confine myself to essentials' (p. 191):

But first I hasten to make a reservation. By and large, popularization is a bad thing. It implants a smattering of learning in society, together with an urge towards the meretricious acquisition of knowledge and indifference to serious and strictly academic work. I'm an enemy of popularization, and in that respect I differ from many eminent scholars, for example Vogt and [*consults a slip of paper*] Moleschott. As recently as last year, I sent each of my aforementioned academic colleagues a letter outlining my view of popularization, but received no answer—probably because I was careless enough to send my letters by ordinary post instead of registering them.

The following other passages were cut out in stage two of revision.

After 'Very well, I commence' (p. 191):

I should have prefaced my lecture with an historical note about when tobacco was first discovered and about the association of ideas whereby humanity came to poison itself with this terrible drug, but, time being short, I must start with the basic essentials.

After '. . . in a quantity strictly corresponding to the number of the girls' (p. 192):

I should have been keeping watch in the kitchen from early morning to dinner time. Never trust servants, ladies and gentlemen! How often the slackness and carelessness of cooks and laundrywomen have caused me to let my wife down! Whenever I've absented myself from the kitchen without permission, the servants have always cashed in on it—thus evoking the just wrath of my patroness. It's true that I've submitted to chastisement with due humility, but the loss occasioned by my carelessness has not been made good. Very well then—today we had pancakes.

After '[. . . *drops his handkerchief on it*]' (p. 192):

Where formulas are concerned, I'm pedantic and inexorable. The girls must remember their formulas as well as their own names.

After '. . . which smells like ammonia' (p. 193):

It is contained in tobacco together with nicotinic acid and nicotianin, which has a pronounced smell of camphor. [*Sneezes.*]

After '. . . but what do we see?' (p. 193):

Take my wife's family, for instance. That family has always marched shoulder to shoulder with my wife's boarding-school, and believe me, not one of my wife's daughters would have permitted herself to behave so thoughtlessly towards an old teacher.

The following passages were removed in stage three of revision.

After 'I even have some of my learned scientific articles printed in the local rag' (p. 191):

For instance, last August saw the publication of my article 'On the Harmfulness of Domestic Animals' under the pseudonym Faust.

After '. . . an 'orrible deadly poison—nicotine' (p. 191):

which is, in my opinion, nothing other than [*looks at his slip of paper and reads out syllable by syllable*]—than volatile nitryl alkaloid-ammonia, where the hydrogen component is replaced by a triatomic radical known to science under the name nicotylene.

After '. . . can rest assured as far as I'm concerned' (p. 192):

so tactful is my wife that the young ladies see in me not so much a being of the opposite sex as a visual aid adapted for the study of that manifestation of higher social organization called the family.

After '. . . which smells like ammonia' (p. 193):

Leaving nicotylene and nicotianin on one side [*sneezes*], let us concentrate our attention on nicotine [*scratches his nose*]. What sort of stuff is it?

(iii) '*Rassokhina's text II*'

This is translated from the text given in *Works*, 1944–51, vol. xii. In that version italics are used to indicate sections of the Russian text which represent handwritten alterations made by Chekhov to the play as lithographed in Rassokhina's second edition. It has been decided not to follow this typographical device in translation. As detailed comparison shows, the last quarter (roughly) of 'Rassokhina's text II' corresponds closely to the text of the final (1903) version. As for the first three-quarters of this penultimate recension, it contains: (i) material peculiar to itself; (ii) material common both to 'Rassokhina's text I' and the final (1903) version; (iii) material common to 'Rassokhina's text I' (but not to the final version); (iv) material common to the final version (but not to 'Rassokhina's text I'). Items (iii) and (iv) account for slightly more material than items (i) and (ii). As these considerations show, 'Rassokhina's text II' is a truly intermediate recension, in view of which Chekhov's contention, that his final (1903) version represented an 'entirely new play', is difficult to accept.

SMOKING IS BAD FOR YOU

A MONOLOGUE IN ONE ACT

(1890, with later alterations of unknown date)

CHARACTER

MARCELLUS NYUKHIN, a hen-pecked husband

The stage represents the platform in the hall of a provincial club

NYUKHIN *struts in majestically. He has long side-whiskers, his upper lip is clean-shaven, and he wears an old, worn tail-coat. He bows and adjusts his waistcoat.*

NYUKHIN. Ladies and, er, in a manner of speaking, gentlemen.

On the supposition that public lectures are, in a manner of speaking, useful, it's been suggested to the wife that I should lecture here in aid of charity on some topic of general interest. I don't see why not.

I'm not a professor, of course, and university degrees have passed me by. Still, it's common knowledge that I've been working for the last thirty years—non-stop, you might even say, ruining my health and all that—on problems of a strictly academic nature. I've even written some learned scientific articles. For instance, I wrote a great screed the other day on 'The Ill Effects Caused by Certain Insects'.

Now, as the subject for my lecture today I've chosen, as it were, the harmful effects of smoking on the human race. On the supposition that it is difficult to explore the full significance of the theme in the scope of a single lecture, I shall endeavour to be brief. As I proceed, I shall, in a manner of speaking, be strictly academic. So, gentlemen, I suggest that you attend with due seriousness to my present lecture. If anyone's scared or put off by the idea of a dry, scientific lecture, he can stop listening and go. [*Makes a majestic gesture and adjusts his waistcoat.*]

I should like to ask the doctors in my audience to pay particular attention. My lecture is a mine of useful information for them, since nicotine not only has harmful effects, but is also used in medicine. For example, we all know about the use of nicotine to exterminate what I might call pernicious insects. Tobacco is, mainly, a plant. [*Hiccups.*] I've got hiccups. A most convenient thing to have too, I might add. It makes you hold your breath and wait a bit. [*A pause of one minute during which* NYUKHIN *stands motionless.*] I've long, er, suffered from hiccups. [*Loses his temper and hiccups.*] That's enough of that, you old clown. This complaint dates from the 13th of September 1889—the very day when my wife gave birth, in a manner of speaking, to our fourth daughter, Veronica. The wife has seven girls in all, but no boys, of which she's very glad because boys would be a nuisance in a girls' boarding-school in many ways. Don't you agree? There's only one man in the whole school—myself. But parents can rest assured as far as I'm concerned.

Actually, time being short, let's not wander from the subject in hand. Now then, where were we? I got hiccups at the most interesting point. Well, never mind. For me and you and especially for the doctors in the house, that incident can serve as an excellent lesson. Between you and me, the fact is that today was pancake day at my wife's boarding-school. The wife runs a school of music, I might add, and a private boarding-school—well, not a boarding-school exactly, but something in that line. Between you and me, the wife has a tidy bit salted away—a cool forty thousand odd—while I haven't a penny to my name, not a bean. Oh well, what's the use of talking? I'm the matron at my wife's boarding-school. I buy food, keep an eye on the servants, do the accounts, make up exercise-books, exterminate bed-bugs, count what you might call the linen, take the wife's dog for walks and catch mice. Last night I had the job of issuing flour and butter to Cook. The pancakes were intended solely for the girls, I might add, while for the members of my wife's family a roast was prepared from a shin of veal which had been in the larder since last Friday. My wife—and I too, of course—decided to proceed on the assumption that it might be no good tomorrow if we didn't cook it today. When the pancakes are already cooked, the wife sends to the kitchen to say that three girls won't get any because they don't want them. It thus transpires that we have several pancakes in hand. What are we to do with them? First my wife wants 'em put in the larder, then she changes her mind. 'You eat 'em, you nitwit!' says she. That's what she calls me when she's in a bad mood—you nitwit. And she always is in a bad mood. Well, I didn't eat the pancakes properly, I just gulped them down because I'm always so hungry.

However [*looks at his watch*], we've somewhat erred and strayed from our subject. So let's go on about tobacco and the terrible harm it causes. Though I've no doubt you'd rather hear a novel or some aria or other. [*Sings.*] 'We'll not be daunted in the heat of battle.' I don't remember how it goes now.

By the way, I forgot to say that besides being matron in the wife's school of music, I also have the job of teaching mathematics, physics, chemistry, geography, history, singing scales, literature and all that. The wife charges extra for dancing, singing and drawing, though I also teach dancing, singing and drawing. Our school of music is at Number Thirteen, Five Dogs Alley in Mrs. Mamashechkin's house—the one whose husband was a major. That's probably why I've always been so nervous—living at Number Thirteen. The wife's at home available to interview parents at any time, and if you want a school prospectus, they're on sale in the porter's lodge at fifty copecks each. [*Takes several prospectuses from his pocket.*] Or I can let you have some of these if you like. Thirty copecks a whack. Any takers? [*Pause.*] None? All right then, make it twenty. [*Pause.*] How annoying! That's it, Number Thirteen.

I'm a complete failure [*looks around him*]. I've grown old and stupid. Here am I lecturing and looking pretty pleased with myself, when I really feel like screaming for help at the top of my voice, as you might say, or taking a run and diving off somewhere head first. [*Gets his snuff-box out of his pocket and takes a pinch.*]

However, I shall speak about snuff and tobacco, and the terrible harm they do to humanity. It's not only men that smoke—women and even girls do it too. They've been mucking around with it again! [*Sneezes.*] Now, what should I do with these wretched, miserable girls? Yesterday they put face-powder in my snuff-box, today it's something burning and poisonous. [*Sneezes and scratches his nose.*] What a dirty trick! [*Almost in tears.*] What a dirty trick, to make fun of an old man like that! [*Sneezes.*] After all, I'm old—I'm sixty-eight. [*Pause.*]

But I notice smiles on many faces. Obviously not all members of the audience fully appreciate the supreme importance of our theme. 'Children,' I always tell my wife's daughters, 'don't laugh at me. After all, you don't know what's going on inside me.' My wife has seven daughters. No, sorry—six, I think. [*Eagerly.*] It's seven! Anna, the eldest, is twenty-seven, and the youngest is seventeen. Gentlemen! [*Looks around him.*] These seven young, unspoilt creatures are an amalgam of everything beautiful, pure and exalted in Nature. At least, that's what my wife tells me to say. Down on my luck I may be, but in fact you see before you the happiest of fathers—I've no choice in the matter, actually, I don't dare say anything else. [*Sighs.*] And not one of them's married yet. [*Shakes his head.*]

Ah me, young men, young men—. By your stubbornness and materialist leanings you deprive yourselves of one of the highest pleasures, that of family life. Though on the other hand, of course, it's better not to get married! If only you knew! Thirty-three years I've lived with my wife—not the best years of my life, perhaps, but not all that bad. In fact they've flashed past like a single moment of ecstasy, actually—blast them!—though I've meekly accepted the punishment which she has inflicted on me. She hasn't yet reached the theatre, she isn't here, so I can tell you everything. [*Looks round and gazes at the wings.*]

The reason why my wife's daughters have been so long finding husbands is probably that they're shy and never meet any men. The wife can't give parties. We can't afford it. They call her a very strict, stingy lady, so people don't come and see us. And it must be admitted, she never has anyone in to a meal, but, er, I can tell you in confidence—. [*Approaches the footlights.*] My wife's daughters are on view on high days and holidays at their Aunt Natalya's—that's the one who has rheumatism and collects old coins. Snacks are served too. And when my wife's away you can get a bit of you-know-what. [*Makes a suitable gesture to indicate drinking.*]

One glass is enough to make me drunk, I might add. It feels good, but indescribably sad at the same time. Somehow the days of my youth come back to me, and I somehow long—more than you can possibly imagine—to escape. [*Carried away.*] To run away, leave everything behind and run away without a backward glance. Where to? Who cares? If only I could escape from this rotten, vulgar, tawdry existence that's turned me into a pathetic old clown! Escape from this stupid, petty, vicious old skinflint of a wife who's made my life a misery for thirty years! Escape from the music, the kitchen, from my wife's doings and stop somewhere in the depths of the country and just stand there like a tree or a post or a scarecrow on some vegetable plot under the broad sky, and watch the quiet, bright moon above you and forget, forget! How I'd love to lose my memory! How I'd love to tear off this rotten old tail-coat that I got married in thirty years ago. [*Takes off his tail-coat.*] The one I always wear when I lecture for charity. So much for you! [*Stamps on the coat.*] Take that! I'm a poor, pathetic old man like this waistcoat with its shabby old back. [*Shows the back.*] I don't need anything! I'm above all these low things. Once I was young and clever and went to college. Now I want nothing—nothing but a bit of peace and quiet. [*Glancing to one side, quickly puts on his tail-coat.*]

I say, my wife's out there in the wings. She's turned up and she's waiting for me there. How simply terrible! [*Looks at his watch.*] But time's up. If she asks, please, please tell her, I beg you, that the lecture was, er—that I behaved with dignity. [*Looking to one side, coughs.*] She's looking this way. [*Speaks, raising his voice.*]

On the supposition, therefore, that tobacco contains a terrible poison, smoking should on no account be indulged in. And I shall venture to hope, in a manner of speaking, that some benefit may accrue from this lecture on 'Smoking is Bad for You'.

That's the end. *Dixi et animam levavi!* [*Bows and struts out majestically.*]

APPENDIX X

THE NIGHT BEFORE THE TRIAL

The play is an unfinished dramatized adaptation of Chekhov's story with the same title [*Ночь перед судом*, 1886]. The present translation is made from the text in *Works*, 1944–51, vol. xii, itself based on the manuscript in Chekhov's hand preserved in the Central State Literary Archive of the U.S.S.R. Calligraphic evidence suggests that the manuscript was written in the middle 1890s.

As readers of this unfinished play can easily deduce for themselves, the denouement would have involved Zaytsev facing Zina's husband as his prosecuting counsel in court next day. Such, at any rate, is the ending of Chekhov's short story with the same title, where the hero comments: 'Looking at him [the prosecutor], I remembered the bugs, little Zina and my diagnosis, whereupon a chill—nay a whole Arctic Ocean—ran down my spine.'

NOTES

The following notes, which have been kept as brief as possible, are designed to explain references in the text which might be obscure to English-speaking readers and to point out certain difficulties which have occurred in the translation.

Page

14. 'Vologda.' Town about three hundred miles north of Moscow.

14. 'St. Tikhon.' Reference is to the Zadonsk Monastery in the Voronezh Province of central Russia, founded in the seventeenth century and famous for its association with St. Tikhon Zadonsky (1724–83).

14. 'Holy Mountains.' A monastery in the Kharkov Province of the Ukraine, said to date from the fifteenth century.

19. 'The Kuban District.' The area of the River Kuban, north of the Caucasus.

28. 'Poltava.' Town in the Ukraine, 70 miles south-west of Kharkov.

29. '. . . as the man says in the play.' Literally, 'as Shchastlivtsev says'. Shchastlivtsev is a character in the play *The Forest* (1871) by A. N. Ostrovsky (1823–86).

40. '*Calchas*.' Though Chekhov does not state the name of the play in which his hero has been acting, this was presumably Shakespeare's *Troilus and Cressida*, the dramatis personae of which include: 'CALCHAS, a Trojan priest taking part with the Greeks.'

43. '*Boris Godunov*.' The passage comes from the verse play *Boris Godunov* (published 1831) by A. S. Pushkin (1799–1837).

43. 'Blow, winds . . . .' From Shakespeare's *King Lear*, Act Three, Scene ii.

43. 'O! the recorders . . . .' From Shakespeare's *Hamlet*, Act Three, Scene ii.

44. 'O silent night . . . .' The passage comes from the second canto of Pushkin's narrative poem *Poltava* (1829).

45. 'Farewell the tranquil mind . . . .' From Shakespeare's *Othello*, Act Three, Scene iii.

45. 'Away from Moscow! . . .' These are Chatsky's famous exit lines from Act Four, Scene xiv of *Woe from Wit* (written 1822–4) by A. S. Griboyedov (1795–1829).

47. 'N. N. Solovtsov.' The famous actor (1856–1902) and childhood friend of Chekhov.

58. 'Whispers, passion's bated breathing.' The first lines of a well-known lyric by A. A. Fet (1820–92).

59. 'Tamara.' A reference to the heroine of the poem *Tamara* (1841) by M. Yu. Lermontov (1814–41).

72. '. . . the goverment land settlement.' Reference is to the statutes issued in connection with the Emancipation of the Serfs in 1861.

77. 'My fate, ye gods . . . .' Famusov's exit lines from the end of Act One of Griboyedov's *Woe from Wit.*

102. 'It's talk like this . . . .' A near-quotation from Famusov's opening speech in Act Two, Scene i of Griboyedov's *Woe from Wit.*

112. 'Oh, tell me not . . . .' From the poem *A Heavy Cross Became her Lot in Life* (1855) by N. A. Nekrasov (1821–78).

117. 'Volunteer Fleet.' This consisted of merchant ships and their crews which could be mobilized in an emergency.

121. 'I loved you once, and love perhaps might still—.' The opening line of a well-known lyric by Pushkin.

122. 'Restless, he seeks the raging storm . . . .' These are the last two lines of Lermontov's well-known lyric *The Sail* (1832).

139. 'Jack the Ripper.' The famous murderer, believed responsible for killing six or more women in London in late 1888. He was never caught, but was believed by some to be a mad Russian doctor.

142. 'Onegin, how can I deny . . . .' From the aria 'To love all ages must submit', sung by Tatyana's husband in the last act of the opera *Eugene Onegin* (1877–88) by P. I. Tchaikovsky (1840–93), based on Pushkin's verse novel with the same title.

149. 'Two friends one night . . . .' From the fable *The Passers-by and the Geese* by I. A. Krylov (1769–1844).

149. 'Oh, tell me not . . . .' See note on p. 112, above.

155. '. . . good old Keating's powder.' Literally, 'Persian powder', a common measure of protection against insects.

164. 'Turgenev.' I. S. Turgenev (1818–83), the well-known Russian novelist.

164. 'Samara.' Town on the middle Volga, now called Kuybyshev.

THE PRONUNCIATION OF RUSSIAN PROPER NAMES

The following is a list of Russian proper names occurring in the text with an indication of their stress, the stressed syllable being indicated by an acute accent over the relevant vowel (see also note to *The Oxford Chekhov*, vol. iii, p. 337).

Aleksándrovich
Altukhóv
Aplómbov
Babelmandébsky
Berezhnítsky
Borís
Bortsóv
Bortsóvka
Cheprakóv
Chubukóv
Dásha
Dmítry
Dýmba
Fédya
Feódorovna
Fínberg
Fyódor
Gánzen
Grendilévsky
Grúzdev
Gúsev
Iván
Ivánov
Kátya
Khamónyev
Kharlámpy
Khírin
Khrápov
Kokóshkin
Korchágin
Kotélnikov
Kubán
Kúritsyn
Kuzmá
Lómov
Maríya
Másha
Matvéyev
Matvéyevich
Mazútov
Mérik
Merchútkin
Mikíshkin
Mirónov
Mísha
Mozgovóy
Muráshkin
Nastásya
Natálya
Natásha
Nazárovna
Nikíta
Nyúkhin
Nyúnin
Odéssa
Olénin
Ólga
Ósip
Patrónnikov
Pelagéya
Polikárpov
Poltáva
Popóv
Répin
Revunóv-Karaúlov
Rýblovo
Sabínin
Samára
Sávva
Sergéy
Shervetsóv
Shipúchin
Smirnóv
Sónnenstein
Sónya
Svetlovídov
Tatyána
Tíkhon
Tolkachóv
Turgénev
Varsonófyevo
Vasíly
Víkhrin
Vlásin
Vlásov
Vólgin
Volódya
Yefímovna
Yegór
Yergóv
Yevdokím
Yevstignéyev
Záytsev
Zhigálov
Zína
Zipunóv
Zmeyúkin

SELECT BIBLIOGRAPHY

I. BIBLIOGRAPHIES IN ENGLISH

Two most useful bibliographies, published by the New York Public Library and containing in all nearly five hundred items, give a comprehensive picture of the literature relating to Chekhov published in English—translations of his writings, biographical and critical studies, memoirs, essays, articles etc. They are:

Chekhov in English: a List of Works by and about him. Compiled by Anna Heifetz. Ed. and with a Foreword by Avrahm Yarmolinsky (New York, 1949) and

The Chekhov Centennial Chekhov in English: a Selective List of Works by and about him, 1949–60. Compiled by Rissa Yachnin (New York, 1960).

Bibliographies in English will also be found in the books by David Magarshack (*Chekhov: a Life*), Ernest J. Simmons and Ronald Hingley mentioned in Section III, below. Magarshack provides a bibliographical index of Chekhov's writings in alphabetical order of their English titles, Simmons includes a list of bibliographies in Russian, and Hingley gives a list of Chekhov's translated stories in chronological order.

II. TRANSLATIONS INTO ENGLISH OF THE PLAYS IN THIS VOLUME

(*a*) IN COLLECTIONS CONTAINING MORE THAN ONE ITEM OF THOSE INCLUDED IN THE PRESENT VOLUME

(Where the titles of translated plays differ from those in the present volume, the title adopted here is given in square brackets if the difference is so great as to make it difficult to identify the play.)

Five Russian Plays. Tr. with an Introduction by C. E. Bechhofer Roberts (London, 1916).

Includes: Chekhov's *The Wedding, The Jubilee* [*The Anniversary*].

Plays by Anton Tchekoff, Second Series: *On the High Road, The Proposal, The Wedding, The Bear, A Tragedian in spite of Himself* [*A Tragic Role*], *The Anniversary, The Three Sisters, The Cherry Orchard.* Tr. with an Introduction by Julius West (New York, 1916).

Plays from the Russian. Tr. Constance Garnett (London, 1923; New York, 1924). Vol. i: *The Cherry Orchard, Uncle Vanya, The Sea-gull, The Bear, The Proposal*. Vol. ii: *Three Sisters, Ivanov, Swan Song, An Unwilling Martyr* [*A Tragic Role*], *The Anniversary, On the High Road, The Wedding.*

Anton Tchekhov: Literary and Theatrical Reminiscences, ed. S. S. Koteliansky (London, 1927).

Includes: *Tatyana Riepin* and *On the Harmfulness of Tobacco* [*Smoking is Bad for You*], tr. Koteliansky.

The Plays of Anton Chekhov: Nine Plays. Tr. Constance Garnett (New York, 1946).

Includes: *The Anniversary, On the High Road, The Wedding, On the Harmfulness of Tobacco* [*Smoking is Bad for You*], *The Bear.*

The Seagull and Other Plays. Tr. with an Introduction by Elisaveta Fen (Harmondsworth, 1954).

Includes: *The Bear, The Proposal, A Jubilee* [*The Anniversary*].

The Brute, and Other Farces by Anton Chekhov. Tr. Eric Bentley and Theodore Hoffman, ed. Bentley (New York, 1958).

Contains: *The Harmfulness of Tobacco* [*Smoking is Bad for You*], *The Brute* [*The Bear*], *Marriage Proposal* [*The Proposal*], *Summer in the Country* [*A Tragic Role*], *A Wedding, The Celebration* [*The Anniversary*].

(*b*) INDIVIDUAL ITEMS

(i) *On the High Road*

On the Highway: a Dramatic Sketch. Tr. David A. Modell, *Drama* (Chicago, 1916), no. 22, pp. 294–322.

(ii) *Swan Song*

Plays by Anton Tchekoff. Tr. with an Introduction by Marian Fell (New York, 1912).

Includes: *The Swan Song.*

(iii) *The Bear*

A Bear. Tr. Roy Temple House (New York, 1909).

The Boor: a Comedy in One Act. Tr. Hilmar Baukhage (New York, 1915).

The Portable Chekhov. Ed. with an Introduction by Avrahm Yarmolinsky (New York, 1947).

Includes: *The Boor.*

(iv) *The Proposal*

The Drama: its History, Literature and Influence on Civilization. By Alfred Bates, 22 vols. (London, 1903).

Includes, in vol. xviii: *A Marriage Proposal.* Tr. W. H. H. Chambers.

A Marriage Proposal: a Comedy in One Act. Tr. Hilmar Baukhage and Barrett H. Clark (New York, 1914).

(v) *Tatyana Repin*

Tatyana Riepin. Tr. S. S. Koteliansky, *London Mercury* (London, 1925), vol. xii, pp. 579–97.

(vi) *A Tragic Role*

The Tragedian in spite of Himself: a Farce of Suburban Life in One Act. Tr. Olive Frances Murphy, *Poet Lore* (Boston, 1922), vol. xxxiii, pp. 268–73.

(vii) *The Anniversary*

The Jubilee: a Farce in One Act. Tr. Olive Frances Murphy, *Poet Lore* (Boston, 1920), vol. xxxi, no. 4, pp. 616–28.

(viii) *Smoking is Bad for You*

The Tobacco Evil. Tr. Henry James Forman, *Theatre Arts Magazine* (New York, 1922), vol. vii, pp. 77–82.

The Harmfulness of Tobacco: a Lecture. Tr. Eric Bentley, *The Caroline Quarterly* (Chapel Hill, 1956), pp. 25–29.

(ix) *The Night before the Trial*

The Night before the Trial. Tr. Anna Heifetz, *The American Mercury* (New York, 1946), vol. lxiii, no. 276, pp. 684–8.

III. BIOGRAPHICAL AND CRITICAL STUDIES

Leon Shestov, *Anton Tchekhov and Other Essays* (Dublin and London, 1916).

William Gerhardi, *Anton Chehov: a Critical Study* (London, 1923).

Oliver Elton, *Chekhov* (The Taylorian Lecture, 1929; Oxford, 1929).

Nina Andronikova Toumanova, *Anton Chekhov: the Voice of Twilight Russia* (London, 1937).

W. H. Bruford, *Chekhov and his Russia: a Sociological Study* (London, 1948).

Ronald Hingley, *Chekhov: a Biographical and Critical Study* (London, 1950).

Irene Nemirovsky, *A Life of Chekhov*. Tr. from the French by Erik de Mauny (London, 1950).

David Magarshack, *Chekhov: a Life* (London, 1952).

David Magarshack, *Chekhov the Dramatist* (London, 1952).

Vladimir Yermilov [Ermilov], *Anton Pavlovich Chekhov, 1860–1904*. Tr. Ivy Litvinov (Moscow, 1956; London, 1957).

W. H. Bruford, *Anton Chekhov* (London, 1957).

T. Eekman, ed., *Anton Chekhov, 1860–1960* (Leiden, 1960).

Beatrice Saunders, *Tchehov the Man* (London, 1960).

Ernest J. Simmons, *Chekhov: a Biography* (Boston, Toronto, 1962; London, 1963).

Maurice Valency, *The Breaking String: the Plays of Anton Chekhov* (New York, 1966).

Thomas Winner, *Chekhov and his Prose* (New York, 1966).

Robert Louis Jackson, ed., *Chekhov: a Collection of Critical Essays* (Englewood Cliffs, N.J., 1967).

Nils Åke Nilsson, *Studies in Čechov's Narrative Technique:* The Steppe *and* The Bishop (Stockholm, 1968).

Karl D. Kramer, *The Chameleon and the Dream: the Image of Reality in Čexov's Stories* (The Hague, 1970).

J. L. Styan, *Chekhov in Performance: a Commentary on the Major Plays* (Cambridge, 1971).

Siegfried Melchinger, *Anton Chekhov*. Tr. by Edith Tarcov (New York, 1972).

Virginia Llewellyn Smith, *Anton Chekhov and the Lady with the Dog*. Foreword by Ronald Hingley (London, 1973).

Harvey Pitcher, *The Chekhov Play: a New Interpretation* (London, 1973).

Sophie Laffitte, *Chekhov, 1860–1904*. Tr. from the French by Moura Budberg and Gordon Latta (London, 1974).

Donald Rayfield, *Chekhov: the Evolution of his Art* (London, 1975).

Caryl Brahms, *Reflections in a Lake: a Study of Chekhov's Four Greatest Plays* (London, 1976).

Ronald Hingley, *A New Life of Anton Chekhov* (London, 1976).

Beverly Hahn, *Chekhov: a Study of the Major Stories and Plays* (Cambridge, 1977).

Kornei Chukovsky, *Chekhov the Man*. Tr. Pauline Rose (London, n.d.).

IV. LETTERS AND MEMOIR MATERIAL ETC.

Letters of Anton Tchehov to his Family and Friends. Tr. Constance Garnett (London, 1920).

The Note-books of Anton Tchekhov together with Reminiscences of Tchekhov by Maxim Gorky. Tr. S. S. Koteliansky and Leonard Woolf (Richmond, 1921).

Letters on the Short Story, the Drama and Other Literary Topics. By Anton Chekhov. Selected and ed. Louis S. Friedland (New York, 1924).

Konstantin Stanislavsky, *My Life in Art*. Tr. J. J. Robbins (London, 1924; New York, 1956).

The Life and Letters of Anton Tchekhov. Tr. and ed. S. S. Koteliansky and Philip Tomlinson (London, 1925).

The Letters of Anton Pavlovitch Tchehov to Olga Leonardovna Knipper. Tr. Constance Garnett (London, 1926).

Anton Tchekhov: Literary and Theatrical Reminiscences. Tr. and ed. S. S. Koteliansky (London, 1927).

Vladimir Nemirovitch-Dantchenko, *My Life in the Russian Theatre*. Tr. John Cournos (London, 1937).

The Personal Papers of Anton Chekhov. Introduction by Matthew Josephson (New York, 1948).

Lydia Avilov, *Chekhov in my Life: a Love Story*. Tr. with an Introduction by David Magarshack (London, 1950).

Konstantin Stanislavsky, *Stanislavsky on the Art of the Stage*. Tr. with an introductory essay on Stanislavsky's 'System' by David Magarshack (London, 1950).

The Selected Letters of Anton Chekhov. Ed. Lillian Hellman, tr. Sidonie Lederer (New York, 1955).

V. OTHER WORKS USED IN THE PREPARATION OF THIS VOLUME

Book of Needs of the Holy Orthodox Church. Done into English by G. V. Shann (London, 1894).

Polnoye sobraniye sochineny i pisem A. P. Chekhova ['Complete Collection of the Works and Letters of A. P. Chekhov'], ed. S. D. Balukhaty, V. P. Potyomkin, N. S. Tikhonov, A. M. Yegolin. 20 vols. (Moscow, 1944–51).

N. I. Gitovich, *Letopis zhizni i tvorchestva A. P. Chekhova* ['Chronicle of the Life and Literary Activity of A. P. Chekhov'] (Moscow, 1955).

Literaturnoye nasledstvo: Chekhov ['Literary Heritage: Chekhov'], ed. V. V. Vinogradov and others (Moscow, 1960).

Chekhov i teatr: pisma, felyetony, sovremenniki o Chekhove-dramaturge ['Chekhov and the Theatre: Letters, Articles and the Comments of Contemporaries on Chekhov as Playwright'], ed. E. D. Surkov (Moscow, 1961).

A. P. Chekhov, *Medved, Predlozheniye, Yubiley* ['*The Bear, The Proposal, The Anniversary*'], with an Introduction, Notes and Vocabulary. Ed. L. M. O'Toole (London, 1963).

A. P. Chekhov, *Svadba* ['*The Wedding*'], with an Introduction, Notes and Vocabulary. Ed. A. B. Murphy (Letchworth, 1963).